Daniela Scheer

The Role of CLCA2 in Tumourigenesis

Daniela Scheer

The Role of CLCA2 in Tumourigenesis

Functional Characterisation of Calcium-Activated Chloride Channel 2 (CLCA2) in Human Tumour Cells

Südwestdeutscher Verlag für Hochschulschriften

Impressum/Imprint (nur für Deutschland/ only for Germany)
Bibliografische Information der Deutschen Nationalbibliothek: Die Deutsche Nationalbibliothek verzeichnet diese Publikation in der Deutschen Nationalbibliografie; detaillierte bibliografische Daten sind im Internet über http://dnb.d-nb.de abrufbar.
 Alle in diesem Buch genannten Marken und Produktnamen unterliegen warenzeichen-, markenoder patentrechtlichem Schutz bzw. sind Warenzeichen oder eingetragene Warenzeichen der jeweiligen Inhaber. Die Wiedergabe von Marken, Produktnamen, Gebrauchsnamen, Handelsnamen, Warenbezeichnungen u.s.w. in diesem Werk berechtigt auch ohne besondere Kennzeichnung nicht zu der Annahme, dass solche Namen im Sinne der Warenzeichen- und Markenschutzgesetzgebung als frei zu betrachten wären und daher von jedermann benutzt werden dürften.

Verlag: Südwestdeutscher Verlag für Hochschulschriften Aktiengesellschaft & Co. KG
Dudweiler Landstr. 99, 66123 Saarbrücken, Deutschland
Telefon +49 681 37 20 271-1, Telefax +49 681 37 20 271-0
Email: info@svh-verlag.de
Zugl.: Wien, Universität Wien, Dissertation, 2009

Herstellung in Deutschland:
Schaltungsdienst Lange o.H.G., Berlin
Books on Demand GmbH, Norderstedt
Reha GmbH, Saarbrücken
Amazon Distribution GmbH, Leipzig
ISBN: 978-3-8381-1200-8

Imprint (only for USA, GB)
Bibliographic information published by the Deutsche Nationalbibliothek: The Deutsche Nationalbibliothek lists this publication in the Deutsche Nationalbibliografie; detailed bibliographic data are available in the Internet at http://dnb.d-nb.de.
 Any brand names and product names mentioned in this book are subject to trademark, brand or patent protection and are trademarks or registered trademarks of their respective holders. The use of brand names, product names, common names, trade names, product descriptions etc. even without a particular marking in this works is in no way to be construed to mean that such names may be regarded as unrestricted in respect of trademark and brand protection legislation and could thus be used by anyone.

Publisher: Südwestdeutscher Verlag für Hochschulschriften Aktiengesellschaft & Co. KG
Dudweiler Landstr. 99, 66123 Saarbrücken, Germany
Phone +49 681 37 20 271-1, Fax +49 681 37 20 271-0
Email: info@svh-verlag.de

Printed in the U.S.A.
Printed in the U.K. by (see last page)
ISBN: 978-3-8381-1200-8

Copyright © 2010 by the author and Südwestdeutscher Verlag für Hochschulschriften Aktiengesellschaft & Co. KG and licensors
All rights reserved. Saarbrücken 2010

The Role of CLCA2 in Tumourigenesis

Functional Characterisation of
Calcium-Activated Chloride Channel 2 (CLCA2)
in Human Tumour Cells

"Grau, teurer Freund, ist alle Theorie. Und grün des Lebens goldner Baum"
Goethe, Faust I (Vers 2038) Mephistopheles

"Als Anwalt würde ich den Krebszellen einfach verbieten zu wachsen"
Alexander Scheer, 30.07.2005

DANKSAGUNGEN

Ich möchte allen die mich bei meiner Dissertation unterstützt haben danken, insbesondere meinen Laborkollegen und meinem Betreuer Univ.-Doz. Dr. Wolfgang Sommergruber, einem großartigen Wissenschaftler und großartigen Menschen.

Besonderer Dank gilt auch meinen Eltern für die Ermöglichung meines Studiums und meinem Mann Alex, der mir Kraft, Ausgeglichenheit, Verständnis und noch viel mehr als nur das gibt.

PREFACE

Due to technical reasons printing of this book was only possible in black and white. As the figures of this PhD-Thesis were originally prepared in colour, some figures may become unclear because of the layout of this book. If you are interested in particular figures in the original layout, please contact me via e-mail at "daniela.scheer@chello.at" and I will send them to you.

TABLE OF CONTENTS

DANKSAGUNGEN 5
PREFACE 7
TABLE OF CONTENTS 9
ABSTRACT 15
ZUSAMMENFASSUNG 17
INTRODUCTION 21
   Cancer 21
   Epithelial and Squamous Cell Carcinoma (SCC) 24
   Identification of the Novel Tumour-Specific Gene CLCA2 26
   The CLCA (CaCC) Family of Putative Calcium-Activated Chloride Channels 26
   Human CLCA2: Present Status and Rationale 27
RESULTS 31
   In Silico Analysis of Expression Data 31
      Expression Profiling of CLCA2 in Normal Tissue Samples 32
      Expression Profiling of CLCA2 in Tumourous Tissue Samples 33
      Expression Profiling of CLCA2 in Tumour Cell Lines 47
      Chromosomal Amplifications of CLCA2 Locus in Tumour Cell Lines 51
      Identification of Differential Gene Expression Profiles 54
   Knock-Down of CLCA2 in CLCA2-Positive Tumour Cells 56
      siRNA Experiments in Human Head and Neck SCC Tumour Cell Line HN5 .. 56
         Evaluation of CLCA2 siRNAs 56
         Proliferation Assay 61
         Apoptosis 62
         Cell Cycle 64
      siRNA Experiments in Human Breast Carcinoma Cell Line MDA-MB-453 65
      siRNA Experiments in a Stable CLCA2-Constitutively-Expressing T47D Clone 69
   Antibody Generation 71

Selection of Peptides for Generation of Antisera 71
Anti-CLCA2 Polyclonal Peptide Antisera and Their Use for Western Blots 73
Immunohistochemistry **74**
Cloning CLCA2 .... **75**
Cloning Various Tags to Expression Vector pCMV-Tag1 75
Amplification of CLCA2 and TA-Cloning to pGEM T-Easy Vector 76
Cloning CLCA2-Tag to Expression Vector pTRE2pur 78
Evaluation of Tags on Western Blot 79
Establishment of Stable wt-CLCA2 Clones **80**
Proteolytic Processing **84**
Mutating Conserved Cysteines 87
Mutational Analysis of the Putative Hydrolase Domain 88
Mutational Analysis of the Putative Cleavage Site 90
Establishment of Stable Mutant-CLCA2 Clones 92
Inhibition of Post-Translational Modifications 94
Integrin β4 .... **97**
Integrin-Mediated Cell Adhesion Assay 98
3D-In Vitro Carcinoma Assay: Spheroids **100**
Growth Curve of CLCA2-Expressing Clones (T47D Tet-on): 2D versus 3D . 107
Affymetrix Microarray Studies **109**
SFI1 113
TP53BP1 114
CDIPT 115
MLL2 116
AFF4 (MCEF) 118
UNC84A 119
DTX2 (Deltex2) 120
C10orf118 121
SATB1 122
EP300 124
LTBP1 125
TSC1 126

TABLE OF CONTENTS

MACF1	128
CHD6	129
RUTBC3	130
P4HA2	131
ASPM	132
Linking Profiles of Cells with Induced CLCA2 Expression to Canonical Pathways	134
Phospho-Proteomic Studies: Canonical Signaling Transduction Pathways	140
Western Blots	141
Antibody Arrays: Proteome Profiler	144

DISCUSSION 147

Introductory Remarks: General Reflections on Targeted Therapy in Oncology	*147*
In Silico Analysis of Expression Data	*150*
Knock-Down of CLCA2 in CLCA2-Positive Tumour Cells	*154*
Antibody Generation and Immunohistochemistry	*155*
Proteolytic Processing	*156*
Putative Hydrolase Domain	156
Putative Cleavage Site	157
Inhibition of Post-Translational Modifications	158
Integrin β4	*158*
3D-In Vitro Carcinoma Assay: Spheroids	*159*
Affymetrix Microarray Studies	*160*
Summary	*164*

MATERIAL AND METHODS 167

Molecular Biology	*167*
Total RNA Isolation	167
Single Strand cDNA Synthesis (First Strand Synthesis)	167
Real-Time Quantitative PCR (RT-qPCR)	168
PCR (Polymerase Chain Reaction)	169
PCR-Amplification	169
Colony Screening	170

TABLE OF CONTENTS

Cloning PCR Products (TA-Cloning) 171
A-Tailing of PCR Products 172
Restriction Digestion 172
Filling 5'-Overhangs with dNTPs 172
Converting a 3'-Overhang to a Blunt End 173
In Vitro Site-Directed Mutagenesis 174
Purification of DNA 175
Phenol/Chloroform Extraction of DNA 175
Precipitation of DNA/RNA 176
Agarose Gel Electrophoresis 176
Recovery of DNA from Agarose Gels 177
Ligation of DNA 177
Transformation of Bacteria 178
    Bacterial Strains 178
    Transformation of JM109 178
    Transformation of JM110 178
Plasmid Preparations 179
    Miniprep 179
    Maxiprep 180
Determination of DNA/RNA Concentration 180
DNA Sequencing 181
Freezing of Bacteria 181

Cell Biology ***182***
Cell Lines 182
    HN5 182
    T47D Tet-on 182
Thawing and Freezing of Cells 182
Propagation of Adherent Cells 183
Determination of Cell Count 183
Determination of Growth Curve 183
Transient Transfection of Plasmid DNA to Eukaryotic Cells 183
Generation of Stable Cell Lines 184

TABLE OF CONTENTS

Generation of Cell Lysates from Human Tumour Cells 185
Inhibition of Post-Translational Modifications 185
Proliferation Assay (Alamar Blue) 185
siRNA Experiments 186
    siRNAs 186
    siRNA Transfection in 6-Well-Plates 186
    siRNA Transfection in 96-Well-Plates 187
    DharmaFECT Reagents 187
Live-Cell High-Content Screening (HCS) with Cellomics ArrayScan® 187
    Tissue Culture 188
    Staining Procedure 188
    Settings 188
    Features 189
Integrin-Mediated Cell Adhesion Assays 189
Human Tissue Samples 190
3D-In Vitro Carcinoma Assay: Spheroids 190
    Methylcellulose 190
    Spheroid Formation 190
    Harvesting of Spheroids 191
    Gel Pouring Device 191
    Collagen Gel Preparation 191
    Matrigel 192
    Probidium Iodide Staining 192
    Measuring of Diameter and Volume of Spheroids 192

Biochemistry ***192***
Determination of Protein Concentration in Cell Lysates (Bradford Assay) ... 192
SDS-PAGE 193
Western Blotting 193
Proteome Analyses with Proteome Profiler^{TM} (R&D Systems) 194
    MAPK-Array 194
    RTK-Array 194
Immunohistochemistry 195

TABLE OF CONTENTS

  Antibody Generation 195
 In Silicio Analyses ***196***
  Statistical and Bioinformatic Analyses 196
  Sequence Alignment 196
  Expression Profiling 196
 Media and Buffers ***197***
  Molecularbiology 197
  Cell Culture 200
 Primer ***201***
  Cloning Primer 201
  Cloning Primer for *In Vitro* Site-Directed Mutagenesis of CLCA2 202
  Sequencing Primers 204
 Oligomers ***205***
 Plasmids ***206***
  pTRE2pur Expression Vector 206
  pGEM-T Easy Vector 206
  pCMV-Tag1 Expression Vector 206
  Human Full-Length cDNA Clone of CLCA2 in pCMV6-XL4 206
 Antibodies ***207***
 APPENDIX **209**
  Sequence of CLCA2 .... ***209***
  DNA Sequence 209
  Potein Sequence 210
  Abbreviations ***215***
 REFERENCES **219**

ABSTRACT

Identification of novel tumour-specific genes and pathways is a key strategy to gain additional prognostic markers and to identify molecular targets for chemical and/or immunological therapy. Combination of subtractive cDNA libraries and cDNA microarrays lead us to identify the human calcium-activated chloride channel 2 (CLCA2) as one gene being highly expressed in lung squamous cell carcinoma (lung SCC) [1]. The presence of T cell epitopes of CLCA2 and its immunological properties have also been described by our group [2]. CLCA2 is a member of the family of calcium-activated cytoplasmic trans-membrane chloride channel proteins [3]. Four human members have been identified, three of them are proteolytically processed into two subunits, whereas CLCA3 represents a truncated and secreted protein [4]. In contrast to other family members, CLCA2 displays a very restricted expression profile in normal tissues (larynx, esophagus, skin).

In the course of this PhD-Thesis a preferential overexpression of CLCA2 in human squamous cell carcinomas (SCC) of different origins, including their lymph node metastases, could be demonstrated. In addition, overexpression was also detected in mycosis fungoides, a cutaneous T cell lymphoma, and in a subset of infiltrating ductal breast cancers. Preliminary IHC analyses confirmed expression of CLCA2 in human SCC samples. *In silico* expression analyses of xenografted SCC cell lines into nude mice revealed a dramatic increase in CLCA2 expression, suggesting a role of CLCA2 in late stage tumourigenesis.

Loss-of-function studies in SCC cell lines by siRNA-mediated knock-down of CLCA2 lead to inhibition of proliferation, reduction in viability, subsequently to induction of apoptotic processes, and to an arrest in G2-phase of the cell cycle. In contrast, breast cancer cell lines stayed unaffected at least in 2D-tissue culture. Inducible expression of CLCA2 in tumour cells in combination with a detailed mutational analysis demonstrated a proteolytic (most likely autocatalytic) processing (hydrolase) activity of CLCA2 and proposed a structural metal site for Zn^{2+} coordination. Altogether, these data strongly support the presence of a zincin-like hydrolase domain in the N-terminal region of CLCA2. As predicted by

ABSTRACT

bioinformatic analyses and in analogy to CLCA1 a monobasic cleavage site could be confirmed by "Alanine-walking" mutagenesis, with the predicted cleavage recognition site: P3(Tyr/Phe) – P2(X) – P1(Arg/Lys) – P1'(Tyr/Phe) – P2'(Phe/Tyr).

Inhibition of glycosylation revealed that both, the stability and the proteolytic processing are impaired when glycosylation is inhibited.

For a better simulation of the situation in a tumour, CLCA2-expressing clones were also grown in multicellular tumour spheroids. In contrast to growth in 2D-cultures a > 2.5-fold increase in the relative proliferation rate in the three-dimensional culture could be identified upon induction of CLCA2 expression.

For a better understanding of the involvement of CLCA2 in signal transduction a comprehensive expression profiling study on Affymetrix GeneChips was performed followed by RT-qPCR and phospho-proteom studies with selected genes. In total ~ 150 genes were shown to be differentially regulated upon induction of CLCA2 expression. Among those genes are CDIPT, MLL2, and SFI1 which play an important role in cell cycle regulation. Utilising Ingenuity Pathways Analysis – a text- and data-mining program – several interaction-networks were identified. All of them could be linked with important canonical signal transduction cascades or interaction networks such as the TNF–IL2–IL10 interaction network or the ERK–MAPK signaling cascade. Phospho-proteom studies confirmed these findings and additionally identified a strong impact of CLCA2 on translational modulation.

Although these expression profiling data can only be regarded as preliminary they, however, offer the basis for further focused studies for the characterisation of CLCA2 as a promising target in SCC tumourigenesis.

ZUSAMMENFASSUNG

Die Identifikation von neuen tumorspezifischen Genen und Signalwegen ist eine Schlüsselstrategie, um neue zusätzliche prognostische Marker und molekulare Targets für chemische und/oder immunologische Therapien zu finden. Eigene Arbeiten führten mittels Kombination von subtraktiven cDNA Banken und cDNA Gen-Chips zur Identifikation des humanen Kalzium-aktivierten Chloridkanals 2 (CLCA2), als ein Gen, das im Plattenepithelkarzinom der Lunge überexprimiert wird [1]. Die Anwesenheit von T-Zell Epitopen und dessen immunologische Eigenschaften wurden ebenfalls von uns beschrieben [2].

CLCA2 ist ein Familienmitglied der Kalzium-aktivierten cytoplasmatischen transmembranen Chloridkanal Proteine [3]. Vier humane Mitglieder wurden bislang identifiziert, drei davon werden proteolytisch in zwei Untereinheiten gespalten, wobei CLCA3 ein trunkiertes segregiertes Protein ist [4]. Im Unterschied zu anderen Familienmitgliedern zeigt CLCA2 ein sehr restriktives Expressionsprofil in Normalgeweben (Larynx, Ösophagus, Haut).

Im Zuge der vorliegenden Dissertation konnte eine präferenzielle Überexpression vom CLCA2 in humanen Plattenepithelkarzinomen verschiedenen Ursprungs, wie auch deren korrespondierender Lymphknoten-Metastasen nachgewiesen werden. Zusätzlich konnte eine Überexpression auch in Mycosis fungoides (einem T-Zell Lymphom, das vorwiegend die Haut betrifft) und in einer Untergruppe von infiltrierenden duktalen Brustkarzinomen gezeigt werden.

Präliminäre immunohistochemische Analysen bestätigten die Expression von CLCA2 in Proben von humanen Plattenepithelkarzinomen. *In silico* Transkriptionsanalysen weisen auf einen dramatischen Anstieg der CLCA2 Expression hin, wenn Plattenepithelkarzinom-Zelllinien in einem Xenograft-Nacktmaus-Modell implantiert werden, was Rückschlüsse auf eine mögliche Rolle von CLCA2 in der späten Phase der Tumorgenese zulässt.

Das Ausschalten der Funktion von CLCA2 in Plattenepithelkarzinomen mittels siRNA führte zur Inhibition der Proliferation, zur Reduktion der Viabilität, zur Induktion von apoptotischen Prozessen und zu einem Arrest in der G2-Phase des

Zellzykluses. Im Vergleich dazu, blieben Brustkarzinom-Zelllinien - zumindest in 2D-Zellkulturen - davon unbeeinflusst. Induzierte Expression von CLCA2 in Tumorzellen in Kombination mit Mutationsanalyse wiesen eine proteolytische (höchstwahrscheinlich autokatalytische) Prozessierungsaktivität (Hydrolaseaktivität) von CLCA2 nach und lassen die Präsenz einer strukturellen Metallionen-Stelle für die Koordination von Zn^{2+} vermuten.

Zusammengefasst weisen diese Daten auf eine Zincin-ähnliche Hydrolasedomäne in der N-terminalen Region von CLCA2 hin.

Wie in bioinformatischen Analysen vorhergesagt und in Analogie zu CLCA1, konnte eine „monobasische Spaltstelle" durch „Alanin-walking" bestätigt werden, wobei die vorhergesagte Spaltstelle sich wie folgt präsentiert: P3(Tyr/Phe) – P2(X) – P1(Arg/Lys) – P1'(Tyr/Phe) – P2'(Phe/Tyr).

Inhibition der Glykosylierung zeigte, dass dadurch sowohl die Stabilität als auch die proteolytische Prozessierung beeinträchtigt werden.

Für eine bessere Simulation der Situation in einem Tumor wurden CLCA2-exprimiernde Klone in multizellulären Tumor-Sphäroiden kultiviert. Im Vergleich zu 2D-Kulturen konnte ein über 2,5-facher Anstieg der relativen Proliferationsrate in der dreidimensionalen Kultur aufgrund von CLCA2 Expression nachgewiesen werden.

Für ein besseres Verständnis der Beteiligung von CLCA2 in der Signaltransduktion, wurde eine umfassende Transkriptionsprofilierungs-Studie mit Affymetrix Gen-Chips durchgeführt. Im Anschluss daran folgten weitere Untersuchungen durch quantitative Real-Time PCR und Phospho-Protein Studien mit ausgewählten Genen. Insgesamt konnte gezeigt werden, dass aufgrund von CLCA2-Induktion ~ 150 Gene differentiell reguliert sind. Unter anderem sind das CDIPT, MLL2 und SFI1, welche eine wichtige Rolle in der Zellzyklus-Regulation spielen. Durch die Verwendung von „Ingenuity Pathways Analysis" – einem „text- and data-mining" – Analysenprogramm konnten Interaktionsnetzwerke identifiziert werden, die allesamt in Verbindung mit wichtigen kanonischen Signaltransduktionskaskaden oder Interaktionsnetzwerken gebracht werden können, wie etwa das TNF–IL2–IL10 Interaktionsnetzwerk, oder die ERK–MAPK Signalkaskade. Phosphoproteom Studien bestätigten in der Folge diese Ergebnisse und konnten zusätzlich den

Zusammenhang dieser Interaktionsnetzwerke mit translationaler Modulation aufzeigen.

Obwohl die Daten der Transkriptionsprofilierung nur ein vorläufiges Zwischenergebnis darstellen, dienen sie als gute Ausgangsbasis für die Planung weiterer Studien zur Charakterisierung von CLCA2 als ein vielversprechendes Target in der Plattenepithel-Tumorgenese.

INTRODUCTION

INTRODUCTION

Cancer

In all industrialised countries, cancer is the second most common cause of death, exceeded only by cardiovascular diseases. To imagine, a total of 1,437,180 new cancer cases and 565,650 deaths from cancer are projected to occur in the United States in 2008 – but these cases do not even include basal cell and squamous cell cancers of the skin [5].

"Cancer" refers to over 150 diseases which have in common the characteristics of uncontrolled growth of cells and the ability to invade and damage normal tissues – locally or at distant sites in the body.

Dependent on the tissue of origin, tumours can be classified into the three major groups of carcinomas, sarcomas and leukaemias. Carcinomas are derived from epithelial cells and account for 90 % of human cancers. Sarcomas are developing from mesodermal cells; they are tumours of the connective or supportive tissue such as bone, cartilage, muscle, blood vessels, fat and of soft tissue. Leukaemias or lymphomas are originating from blood or the lymphatic system, respectively. Furthermore, there is the group of melanomas, which are malignant tumours of melanocytes, which are found predominantly in skin and gliomas, which are tumours with its origin in brain or spine (nerve tissue).

Cancer is predominately a genetic disease arising from the accumulation of mutations [6], either inherited (germline) or acquired (somatic), in critical proto-oncogenes or tumour suppressor genes. Also epigenetic changes can cause cancer [7], such as loss of imprinting [8] or hyper- or hypomethylation of 5-methylcytosine [9], as well as chromosomal instability [10] and microsatellite instability [11].

The transformation of a normal cell into a highly malignant tumour cell – the process of tumourigenesis – is a multistep process [12-14] in terms of stepwise genetic variation. The originally proposed model of Nowell [12], suggests the evolution of tumour cell populations in terms of stepwise genetic variation by an induced change in a single previously normal cell which makes it neoplastic and

INTRODUCTION

provides it with a selective growth advantage over adjacent normal cells. As a result of genetic instability in the expanding population, mutant cells are produced, which are nearly all eliminated due to metabolic disadvantages or immunologic destruction, but occasionally one has an additional selective advantage and this mutation becomes the precursor of a new predominant subpopulation. Over time there is a sequential selection by an evolutionary process of sublines which are genetically and biologically increasingly abnormal. Different tumours acquire certain similarities, but divergences occur as well, as local conditions in each neoplasm differently affect the emergence of variant sublines.

Frequently, a single cancer cell contains mutations in a lot of genes, gross chromosomal abnormalities and widespread changes in its gene expression profile [15]. Hanahan and Weinberg suggest a succession of six independent steps of essential alteration in the genotype of a cell, whereby each step represents a physiological barrier that must be overcome in order for a cell to progress further toward the end point of malignancy [13;16]. These steps are:

- *"Self-sufficiency in growth signals"*, which will be a first step towards loss of growth regulation in a normal cell. The tumour cell becomes independent of exogenous growth stimuli, as e.g. the tumour cell produces their own growth factors in an autokrine mechanism. As an example, PDGF (platelet-derived growth factor) and TNFα (tumour necrosis factor α) is produced by glioblastomas and sarcomas [17].

- *"Insensitivity to antigrowth-signals"*: Mutations in tumour suppressor genes may induce the proliferative potential of a tumour cell, when the tumour suppressor gene normally protects cells of uncontrolled proliferation by regulation of the transition of G1- to S-phase in the cell cycle and normally force cells into the quiescent (G0) state. These signals are mainly transduced by soluble growth inhibitors or inhibitors embedded in the extracellular matrix (ECM) or located on the surface of neighbouring cells. Tumour cells often interfere with the Rb (retinoblastoma) tumour suppressor, which controls most of the antiproliferative signals in the cell, by mutation or hypophosphorylation [18;19]. Alternatively, cells may be induced to permanently relinquish their proliferative potential by being induced to enter into postmitotic states, usually

INTRODUCTION

associated with acquisition of specific differentiation-associated traits. Another way to avoid the antiproliferating state is the downregulation of integrins which mediate various intracellular signals. By this, integrins not only define cellular shape and mobility but also regulate the cell cycle. Integrins essentially contribute to cell division, cell survival, cellular differentiation and apoptosis.

- *"Evasion of apoptosis"* which means that cancer cells acquires resistance towards apoptosis and do not undergo apoptosis, if the cells have been damaged. Therefore, tumour cells disrupt sensors for apoptosis by e.g. mutations in the tumour suppressor gene p53, which affords unlimited replication. On the other hand downregulation of Fas receptors and killing of activated T-lymphocytes through the constitutive expression of Fas-ligand by tumour cells has been suggested as a mechanism for pathologic suppression of immune surveillance [19], and has been demonstrated in melanomas and colon cancers as an important antiapoptotic mechanism [20;21].

- *"Limitness of replicative lifespan"* can be obtained in a tumour cell by upregulation of e.g. the enzyme telomerase, which maintains the lengh of telomeres [22] and therefore contributes to the capability of unlimited replication.

- *"Sustained angiogenesis"* is achieved by changing the balance of angiogenic inducers and inhibitors to induce and sustain angiogenesis [23] e.g. by increased expression of proangiogenic factors such as vascular endothelial growth factor (VEGF) and/or fibroblast growth factor (FGF).

- *"Tissue invasion and metastasis"*: To transform a primary tumour into an invasive and metastasising tumour, which is able to invade local tissues and blood as well as lymphatic vessels, changes in the expression pattern of adhesion molecules like cell-cell adhesion molecules (CAM's) and integrines are necessary to make the tumour less adhesive [24]. Furthermore, upregulation of extracellular matrix-degrading proteases supports the evasion of tumour cells through the basal cell membrane across blood vessel walls and into connective tissues. Such a cell-to-cell interaction molecule is e.g. E-cadherine, which is also involved in β-catenin signalling [25]. Recently, it was shown that expression of E-cadherin correlates with the presence of nodal metastases at the time of diagnosis in SCC of the head and neck [26;27].

This very general overview on cancer demonstrates that the cause of cancer is not linked to a single molecular event but in contrast, tumourigenesis encompasses complex molecular interactions and deregulation of signaling cascades in which single proteins are involved. Some of these altered proteins may play an essential role, others may only be regarded as passenger alterations (mutations).

"Oncogene addiction" describes the acquired dependence of tumour cells on an activated oncogene for their survival and/or proliferation. This phenomenon has important implications for the success of targeted cancer therapies [28;29]. It was first discussed for mutations in the EGFR gene of human NSCLCs [30;31].

Epithelial and Squamous Cell Carcinoma (SCC)

An epithelium is defined as a layer of cells that is located at the boundary of body cavities and the connective tissues. Epithelial cells are arranged along the basement membrane, which mainly consists of collagens. As shown in Figure 1 different types of epithelial structures are classified by shape, stratification and specialisation. Squamous epithelium can be distinguished by two types of "simple squamous epithelia", which is single-layered, and the type of stratified squamous epithelia, which is multilayered (Figure 1). Squamous epithelial exhibits a typically flat and scale-like shape. Several molecular biomarkers for squamous epithelium were identified, which are human desmocollin 3 (DSC3) [32], keratin 5 (KRT5) [33], desmoglein 3 (DSG3) [34], plakophilin 1 (PKP1) [35], plakophilin 3 (PKP3) [36], PERP [37], and syndecan [38].

Squamous cell carcinoma (SCC) specifically derives from squamous epithelial cells and may occur in many different organs, including skin, lung, cervix, prostate, esophagus, urinary bladder, and the head and neck region.

Accounting for approximately 20 % of cutaneous malignancies, SCC is the second leading cause of skin cancer in Caucasians [39] (after basal cell carcinoma, but more common than melanoma) with over 250,000 new cases per year estimated in the United States, whereby incidence appears to be in the rise [40]. Most cases of head and neck cancer are due to SCC. About 30 % of all lung cancers are SCCs.

Two thirds of all SCCs are located in the central lung, the remaining one third is peripheral.

In the carcinoma in situ (the earliest form), the squamous epithelium is replaced by malignant squamous cells and with ongoing growth, tumour cells invade the sub-epithelial basement membrane. Carcinomas in situ do not necessarily progress to invasive carcinomas, but they can spontaneously regress to normal mucosa. The histological pattern of SCC appears as intracellular bridging, horn pearl formation and individual cell keratinisation; these features become more and more characteristic, the higher such a tumour is differentiated. About 5 % of SCCs are invasive and metastasise to different sites in the body [41]. For most of the SCCs there exists a high medical need for novel targeted therapies. Especially in recurrent and/or metastatic SCC of the head and neck, there have been no improvements in survival in last decade [42]. Therefore, the identification of SCC-specific proteins is a key element to gain additional prognostic markers and a prerequisite to develop novel chemical and immunological therapeutic strategies for treatment of SCCs.

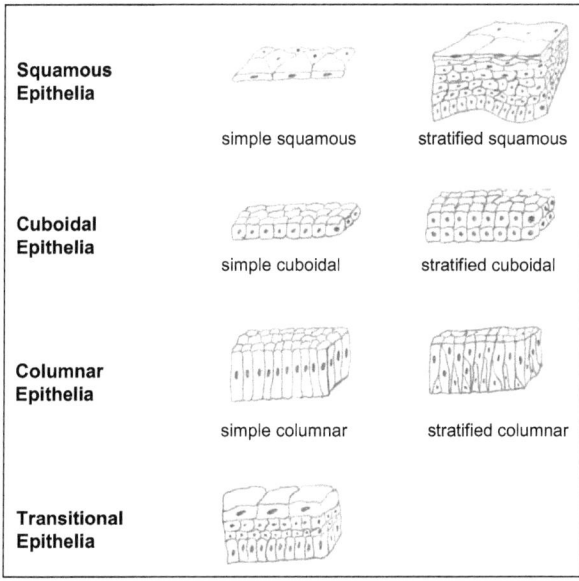

Figure 1. Different types of epithelia

Identification of the Novel Tumour-Specific Gene CLCA2

Combination of subtractive hybridisation using a pool of lung squamous cell carcinoma (lung-SCC) as tester and a pool of 11 critical normal tissues as driver on cDNA microarrays lead to the identification of the human calcium-activated chloride channel 2 (CLCA2) as one gene being highly expressed in lung-SCC [1]. In addition, selected HLA-A2-restricted peptides from CLCA2 have been selected and by using *in vitro* priming, T cell lines against the CLCA2-derived KLLGNCLPTV, LLGNCLPTV, and SLQALKVTV peptides have been generated. These primed T cells also recognised allogeneic tumour cells in an antigen-specific and HLA-restricted fashion. Moreover, peptide LLGNCLPTV was also independently recognised by CD8(+) T cells expanded from pancreatic carcinoma/T cell cocultures whereas CLCA2-specific CD8(+) T cells were absent from the peripheral blood of healthy donors. These data indicate that an immune response can be induced against CLCA2, which thus may become an important antigen for anti-tumour vaccination approaches [2]. However, the contribution of CLCA2 in tumour growth and/or survival and its biological role in tumourigenesis is still unknown.

The CLCA (CaCC) Family of Putative Calcium-Activated Chloride Channels

The CLCA gene family is a growing family of trans-membrane proteins with a putative role in chloride conductivity across the outer cell membrane, that is regulated by the intracellular concentration of calcium [3]. CLCA proteins are of high interest for respiratory disorders with chronic mucus overproduction, including asthma, chronic obstructive pulmonary disease (COPD), and cystic fibrosis.

To date, 17 homologous genes have been identified so far in mammalians. The family consists of the two bovine founder molecules: bCLCA1 (alias CaCC) [43;44] and bCLCA2 (alias Lu-ECAM-1) [45;46]; six murine homologues: mCLCA1 [47;48], mCLCA2 [49], mCLCA3 (alias gob-5) [50], mCLCA4 [51], mCLCA5 [52] and mCLCA6 [52]; four human homologues: hCLCA1 [53;54], hCLCA2 [55;56], hCLCA3 [4], hCLCA4 (CaCC2) [54]; one porcine homologue: pCLCA1 [57]; a rat homologue:

rCLCA1 [58] and two rat-brain homologues: rbCLCA1 [59] and rbCLCA2 [60]; one equine homologue: eCLCA1 [61]; and one canine homologue: cCLCA1 [62].

All members of this gene family map to the same site on chromosome 1p31-p22 [63] and share high degrees of homology in size, sequence, and predicted structure; but they differ significantly in their tissue distribution and in cellular expression pattern.

With the exception of the truncated secreted member human CLCA3 [64], all are synthesised as an 902-943 amino acid precursor trans-membrane glycoprotein of approximately 125 kDa weight, which is rapidly cleaved into 90 and 35 kDa subunits [4]. Another common feature is a symmetrical multiple Cysteine motif in the amino-terminal tail. For human CLCA1 a novel conserved metallo-hydrolase structural domain was shown [65].

Despite the well characterised biophysical property as chloride channels, demonstrated by cell transfection and patch clamp studies, other functions and their physiological significance are poorly understood. Independently of their alleged role in ion conductance, the field of functionalities seems to be quite broad. Certain CLCA family members may serve as cell-adhesion molecules and may play a role in tumour metastasis. Aberrant morphology of cells overexpressing murine CLCA2 has been interpreted as evidence for modulation of the cell cycle by CLCA proteins [66]. A mechanism of tumour suppressive effects of CLCA proteins has been reported, but is not understood in detail yet.

Human CLCA2: Present Status and Rationale

The human CLCA2 protein is highly homologous to murine CLCA5 [52]. It is expressed as a 943 amino acid precursor whose N-terminal signal sequence is removed, followed by internal cleavage near amino acid position 670. Biochemical analysis of human CLCA2 suggested five trans-membrane domains [64], three of which were predicted to be located in the larger amino-terminal part and two in the smaller ~ 38 kD carboxy-terminal cleavage product. More recent investigations of trans-membrane geometry predict that a C-terminal 22-amino-acid hydrophobic segment comprises the only trans-membrane pass [67]. However, the native structure remains unclear.

A tumour suppressive effect of human CLCA2 has been suggested through pro-apoptotic action: Stable transfection of CLCA2 into a human breast carcinoma cell line reduced its invasive growth *in vitro* and slowed their tumourigenicity *in vivo* [66;68]. The von Willebrand factor type A (vWA) domain has been discovered in CLCA2 [69], and was shown to play a role in adhesion by being involved in the binding of Integrin β4 to human CLCA2.

In contrast to other family members, human CLCA2 displays a very restricted tissue expression profile. It is selectively expressed in human lung, trachea and mammary gland [55]. Expression in uterus, prostate, testis was also reported [54], as well as in corneal epithelium [56]. Since this protein is expressed predominantly in trachea and lung, it is suggested that CLCA2 might play a role in the complex pathogenesis of cystic fibrosis. It may also serve as adhesion molecule for lung metastatic cancer cells, mediating vascular arrest, and colonisation.

In a tumour attachment model the expression of human CLCA2 on lung endothelial cells was demonstrated to mediate the attachment of breast cancer cells, by binding to Integrin β4 [70]. The additional function of this family of proteins as adhesion molecules is supported by the prediction of a von Willebrand factor type A adhesion domain [71]. This domain was first described as the adhesion element in blood-clotting proteins. Binding between murine CLCA1 and Integrin β4 was demonstrated to induce the activation of the focal adhesion kinase in cancer cells [72]. The adhesion function of this class of proteins was further demonstrated by the fact that the lung epithelial bovine CLCA2 was shown to mediate lung metastasis of melanoma cells in mice [73].

Multiple tissue Northern blot analysis revealed a very restricted expression profile in normal tissues as it was only detected in larynx and esophagus. In addition, a weak expression was detected in skin and testis (PhD-Thesis of Ulrich König 2001, Boehringer Ingelheim Austria). These findings are supported by the work from Pauli and Gruber [54;55]. *In situ* hybridisations and a large set of comparative hybridisations on cDNA arrays followed by real time PCR studies supported this result; 100 % of lung SCC patient samples exhibit high upregulation of human CLCA2, whereas residual normal lung was negative for human CLCA2.

INTRODUCTION

Suggesting CLCA2 as a novel marker for SCC, these data served as the basis for the present PhD-Thesis, with the aim of the functional characterisation of CLCA2 in human tumour cells. Recent data of gene expression profiling of primary and metastatic melanoma reported differential gene expression of human CLCA2 between metastatic melanoma and non-metastatic cutaneous tumours [74], and confirmed data which have been obtained during this PhD-Thesis.

Based on previous studies of our group and others (see above), there is profound evidence that CLCA2 is a tumour-associated protein which might play an important role in sqamous cell carcinogenesis. However, very little is known about the function of CLCA2 and its essential contribution to tumour cell growth and/or survival.

Therefore, the aims of this PhD-Thesis can be summarised as follows:

- performing a detailed *in silico* expression profile analysis utilising mainly the in-house expression database BioExpress (GeneLogic Inc.),
- functional characterisation of CLCA2 by gain- and loss-of-function studies in selected SCC cell lines,
- identification of functionally essential amino acids of CLCA2 by mutational analyses (putative metallo-hydrolase domain),
- analysing the effect of wild type and mutant CLCA2 in 3D culture,
- expression profiling on Affymetrix GeneChips of cell lines upon induction with wild type and mutated CLCA2, selecting differentially induced genes and analysing those genes by RT-qPCR and Western blot analysis, and finally
- utilising data mining algorithms to bring CLCA2 and the differentially induced genes into a biological context.

RESULTS

RESULTS

This PhD-Thesis is based on data of previous projects in the Sommergruber lab at Boehringer Ingelheim Austria, which aimed at the identification of novel putative targets for anti-cancer therapy. In these studies, Stefan Amatschek compared during his diploma thesis tumourous and non-tumourous patient samples by analysing 20 breast carcinomas, 11 lung squamous cell carcinomas (SCC), 11 lung adenocarcinomas (AC), 8 renal cell arcinomas (RCC), and 16 corresponding normal tissues utilising PCR-based cDNA subtraction analyses combined with subsequent cDNA microarray analyses. The obtained data allowed the discrimination between mRNA transcripts, which are exclusively expressed in lung SCC and/or lung ACs [1]. One of these transcripts was identified as the human calcium-activated chloride channel, family member 2 (CLCA2).

During his PhD-Thesis, Ulrich König could confirm the lung SCC-specific mRNA expression of CLCA2 by different methods such as Northern blot, RT-qPCR, *in situ* hybridisation and others. Additionally, he could identify an overexpression of CLCA2 mRNA transcript in esophageal SCCs in comparison to normal esophageal tissue. In addition, he could show that selected HLA-A2-restricted peptides from CLCA2 can be used for *in vitro* priming T cell lines [2].

In Silico Analysis of Expression Data

The expression levels of CLCA2 mRNA were investigated in available normal tissue samples, tumourous tissue samples and tumour cell lines utilising Boehringer Ingelheims proprietary BioExpress database (GeneLogic Inc.), which includes expression data from ~ 14.000 human tissue samples, analysed on Affymetrix GeneChips.

Expression Profiling of CLCA2 in Normal Tissue Samples

By comparison of 31 different normal tissues using BioExpress database (GeneLogic Inc.), CLCA2 expression could only be found in normal skin, larynx and esophagus (Figure 2).

The expression of CLCA2 mRNA seems to be tightly regulated, as its mRNA is only found in a small number of normal tissues and is not ubiquitously expressed.

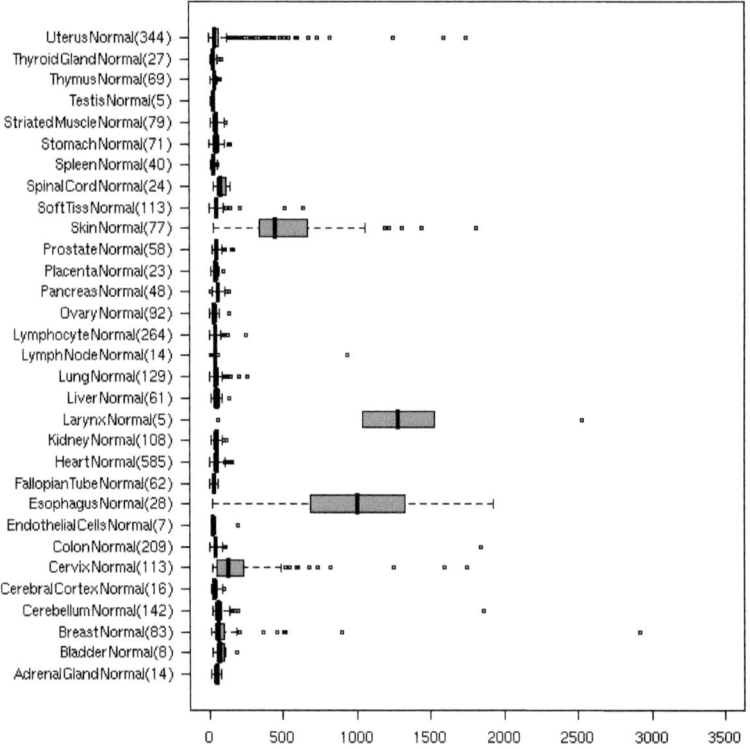

Figure 2. CLCA2 mRNA expression in normal tissue samples
CLCA2 mRNA expression was analysed using Affymetrix GeneChip technology. The Affymetrix identification code for CLCA2 is 206165_s_at. The expression levels of CLCA2 mRNA are shown in box-plots. The vertical centre line in the box indicates the median. The box itself represents the interquartile range (IQR) between the first and third quartiles. Whiskers extend to 1.5 times the IQR; the position of extreme values, if above the upper whisker limits, are marked by open circles. The human sample collection has been described by the originator of the BioExpress database (GeneLogic Inc.). Numbers in brackets indicate the sample numbers. All samples have been MAS5-normalised [75].

Expression Profiling of CLCA2 in Tumourous Tissue Samples

The expression levels of CLCA2 were examined in those available tumourous and corresponding normal tissue samples, where the disease annotation was confirmed by the revision of the clinical history by an experienced pathologist. The verified diagnostic sample annotation was taken as the primary quality criteria for sample selection.

Varying portions of tumour cells on the individual tissue sections could distort some of the relative expression values. In-depth statistical analyses of those samples, of which the percentage of tumour cells per section was available and accordingly correlated (data not shown), resulted in negligible corrections of the relative expression values.

Data for cervix carcinomas could only be linked to squamous cell carcinomas (SCC). For those two indications no further tumour type sub-classification could be found in BioExpress database (GeneLogic Inc.). Tumours from the lung were subdivided into small cell lung cancers (SCLC), adenocarcinomas (AC), adenosquamous carcinomas (AdSqCa), large cell carcinomas (LCC), and squamous cell carcinomas (SCC). For a fraction of SCCs from the head and neck region it was possible to classify those samples into two sub-categories (esophageal and laryngeal tumours), since data of the corresponding normal tissue were also available. For tumours of the urinary bladder mainly data on transitional cell carcinomas (TCC), the most common type of bladder carcinomas, were available. Lymph node metastases were subcategorised, corresponding to the tumour subtype of the primary tumour they descend from, namely melanoma (Mel), adenocarcinomas (AC), and squamous cell carcinomas (SCC).

As previously mentioned, CLCA2 expression was already detected in esophageal and lung SCCs by Northern blotting and RT-qPCR. As shown in Figure 4, analyses of the expression pattern in the BioExpress database (GeneLogic Inc.) confirmed these data and indicated a preferential overexpression of CLCA2 for lymph node metastases of SCCs (LNSCCMet) as well as for SCCs of the larynx, the cervix, the head and neck region, and of a subpopulation of the infiltrating ductal breast carcinoma and mycosis fungoides.

RESULTS

There is a clear preference of elevated levels of CLCA2 mRNA transcript in SCCs such as larynx SCC, head and neck SCC or cervix SCC (Figure 4).

After in-depth analyses of the available clinical history of the normal skin samples a discrimination between skin samples derived from healthy individuals and skin samples from cancer patients was possible. As the expression levels for CLCA2 are significantly higher in most of the skin samples, which were taken from cancer patients distant from the primary tumour site (Figure 3), it is very likely that the expression levels of CLCA2 in tumour-associated normal samples of larynx and esophagus are similarly altered by distant primary tumours. However, for these tissues samples no reliable clinical history was available.

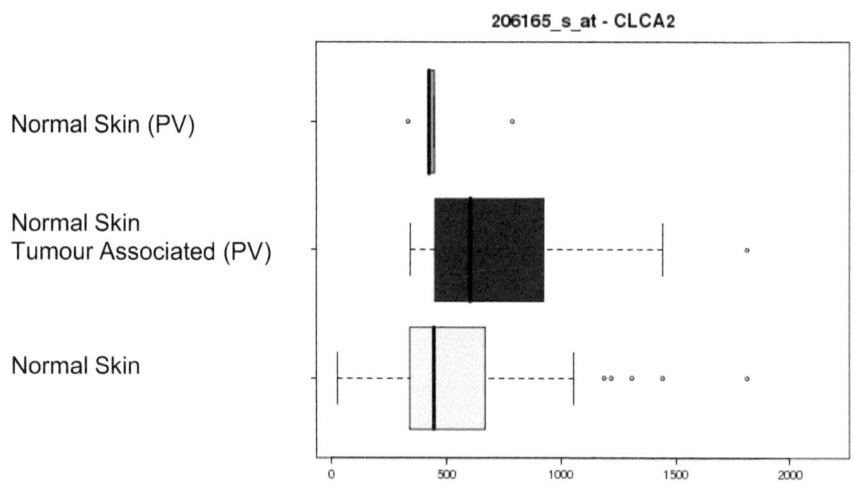

Figure 3. CLCA2 mRNA expression in skin samples of different origins
PV: Verified by pathologist; above: Skin-derived from non-tumour (normal) samples; middle: Normal skin tumour-associated samples; below: Normal skin samples w/o assignment. CLCA2 mRNA expression was analysed using Affymetrix GeneChip technology. The Affymetrix identification code for CLCA2 is 206165_s_at. The expression levels of CLCA2 mRNA are shown in box-plots. The vertical centre line in the box indicates the median. The box itself represents the interquartile range (IQR) between the first and third quartiles. Whiskers extend to 1.5 times the IQR; the position of extreme values, if above the upper whisker limits, are marked by open circles. The human sample collection has been described by the originator of the BioExpress database (GeneLogic Inc.). Numbers in brackets indicate the sample numbers. All samples have been MAS5-normalised.

In contrast to SCCs, overexpression of CLCA2 is neither detected in ACs and their corresponding lymph node metastasis, nor in tumours of the hematopoietic or lymphatic system indicating again its highly SCC-specific expression (Figure 4).

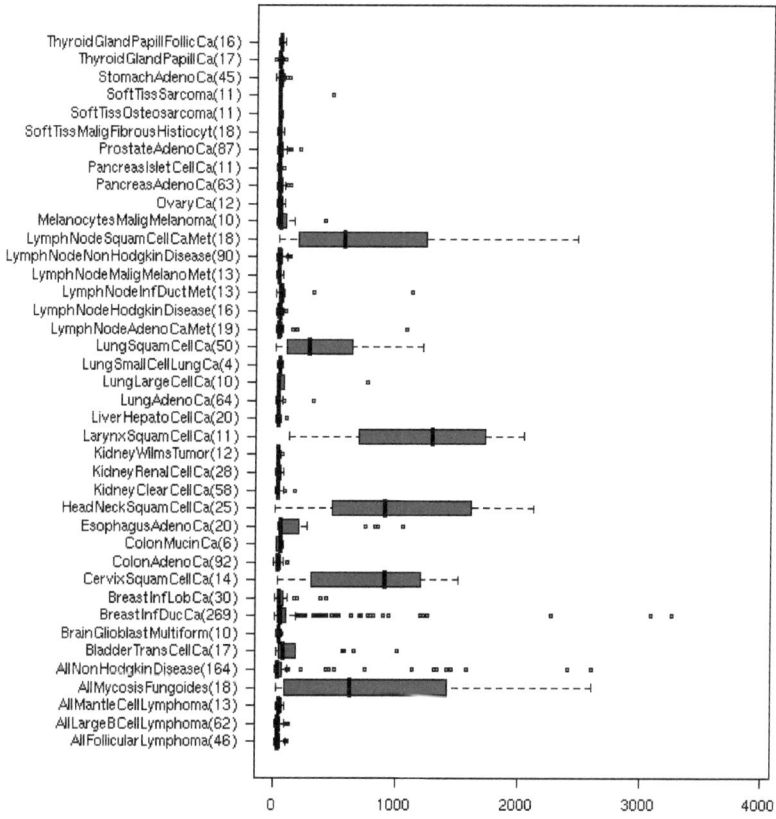

Figure 4. CLCA2 mRNA expression in various tumours
All: Acute lymphocytic leukaemia, Ca: Carcinoma, Met: Metastases, Tiss: Tissue, Squam: Sqamous, InfDuc: Infiltrating ductal, InfLob: Infiltrating lobular, PapillFollic: Papillary and follicular, Malig: Malignant, TransCellCa: Trans cell carcinoma, ColonMucinCa: Mucinous colon carcinoma, GlioblastMultiform: Glioblastoma multiforme.
CLCA2 mRNA expression was analysed using Affymetrix GeneChip technology. The Affymetrix identification code for CLCA2 is 206165_s_at. The expression levels of CLCA2 mRNA are shown in box-plots. The vertical centre line in the box indicates the median. The box itself represents the interquartile range (IQR) between the first and third quartiles. Whiskers extend to 1.5 times the IQR; the position of extreme values, if above the upper whisker limits, are marked by open circles. The human sample collection has been described by the originator of the BioExpress database (GeneLogic Inc.). Numbers in brackets indicate the sample numbers. All samples have been MAS5-normalised.

Figure 5 comprises a comparison of the major therapeutically relevant tumour types versus their corresponding normal tissues. Elevated levels of CLCA2 mRNA transcript were detected only in lung SCC and could not be found in lung AC or the corresponding normal tissues. In prostate and colon carcinomas no CLCA2 expression was found (Figure 5).

In breast cancer, a subpopulation of the infiltrating ductal carcinomas shows overexpression of CLCA2, which is represented by the high amount of outliers in Figure 5. However, it remains unclear whether distinct chromosomal aberrations or different differentiation stages lead to the induction of and/or dependence on CLCA2 in this subset of breast cancer samples. In this context it has to be mentioned that Her2 amplification is also found only in a subset (outlier) of breast cancer samples.

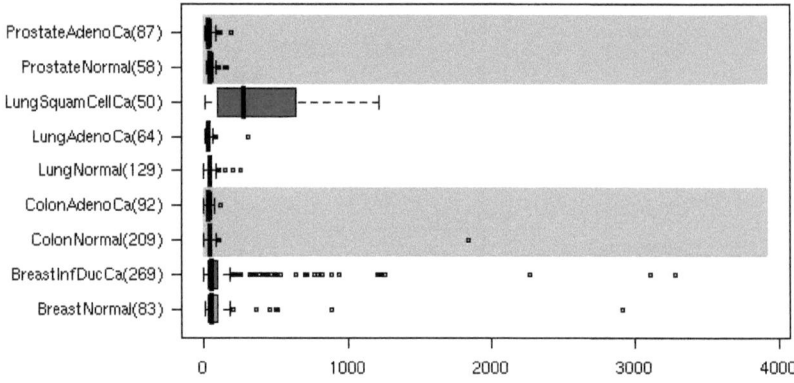

Figure 5. CLCA2 mRNA expression in tumourous tissues: Major therapeutically important tumour types
CLCA2 mRNA expression was analysed using Affymetrix GeneChip technology. The Affymetrix identification code for CLCA2 is 206165_s_at. The expression levels of CLCA2 mRNA are shown in box-plots. The vertical centre line in the box indicates the median. The box itself represents the interquartile range (IQR) between the first and third quartiles. Whiskers extend to 1.5 times the IQR; the position of extreme values, if above the upper whisker limits, are marked by open circles. The human sample collection has been described by the originator of the BioExpress database (GeneLogic Inc.). Numbers in brackets indicate the sample numbers. All samples have been MAS5-normalised. Abbreviations see Figure 4.

RESULTS

In Figure 6 lymph node metastases are compared with a set of "minor indications" and their corresponding normal tissues. Again, CLCA2 overexpression could only be detected in lymph node metastases derived from SCC but not AC-driven metastases.

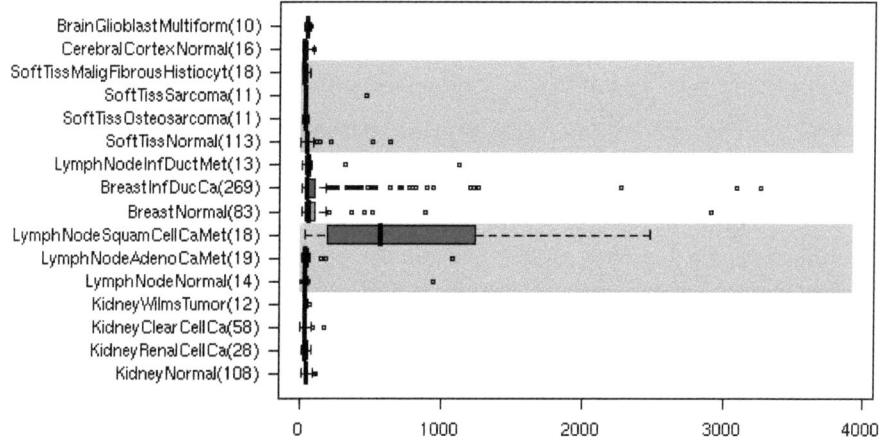

Figure 6. CLCA2 mRNA expression in lymphnode metastases
SoftTissMaligFibrousHistiocyt: Soft tissue malignant fibrous histiocytoma; other abbreviations see Figure 4.
CLCA2 mRNA expression was analysed using Affymetrix GeneChip technology. The Affymetrix identification code for CLCA2 is 206165_s_at. The expression levels of CLCA2 mRNA are shown in box-plots. The vertical centre line in the box indicates the median. The box itself represents the interquartile range (IQR) between the first and third quartiles. Whiskers extend to 1.5 times the IQR; the position of extreme values, if above the upper whisker limits, are marked by open circles. The human sample collection has been described by the originator of the BioExpress database (GeneLogic Inc.). Numbers in brackets indicate the sample numbers. All samples have been MAS5-normalised.

In the view of elevated CLCA2 transcript levels in tumour-associated skin biopsies (see Figure 3) it is of interest that a dramatic increase of CLCA2 mRNA was identified in mycosis fungoides (Figure 7). Mycosis fungoides is the most common type of cutaneous T cell lymphoma (CTCL). It is a Non-Hodgkin lymphoma, which is predominantly related to skin. It generally affects the skin, but may progress internally over time. As shown in Figure 3, mRNA expression of CLCA2 seems to be induced in tumour-associated skin rather than in normal skin. Similarly, mycosis fungoides might make use of CLCA2 for its colonisation of the skin during its course of progression. In contrast, no CLCA2 expression is found in Hodgkin lymphomas.

In summary, it seems that formation of SCC-derived metastases are somehow linked with CLCA2 expression (Figure 6) supporting the idea that CLCA2 might play a crucial role in the formation of SCC-derived metastases and for the progression of mycosis fungoides (Figure 7).

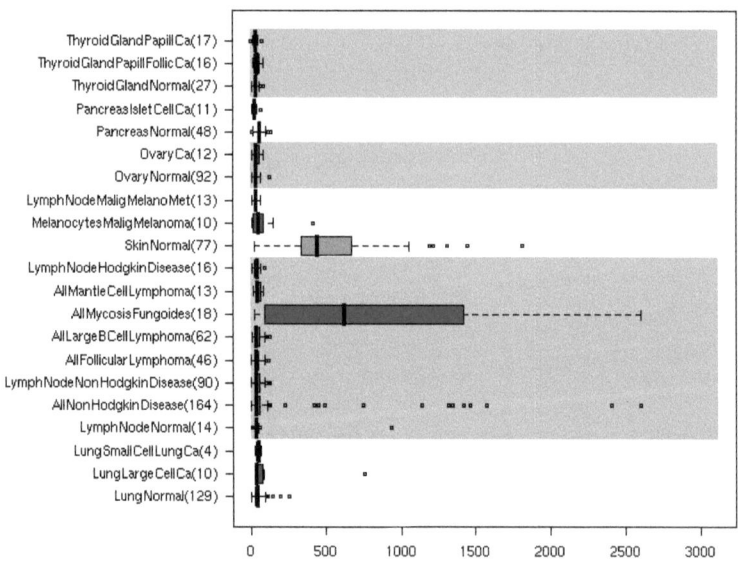

Figure 7. CLCA2 mRNA expression in lymphomas and mycosis fungoides
CLCA2 mRNA expression was analysed using Affymetrix GeneChip technology. The Affymetrix identification code for CLCA2 is 206165_s_at. The expression levels of CLCA2 mRNA are shown in box-plots. The vertical centre line in the box indicates the median. The box itself represents the interquartile range (IQR) between the first and third quartiles. Whiskers extend to 1.5 times the IQR; the position of extreme values, if above the upper whisker limits, are marked by open circles. The human sample collection has been described by the originator of the BioExpress database (GeneLogic Inc.). Numbers in brackets indicate the sample numbers. All samples have been MAS5-normalised. Abbreviations see Figure 4.

RESULTS

In contrast to mycosis fungoides no (elevated) expression of CLCA2 was detectable in B cell and T cell lymphomas and their corresponding normal tissues/cells (Figure 8, Figure 9 and Figure 10).

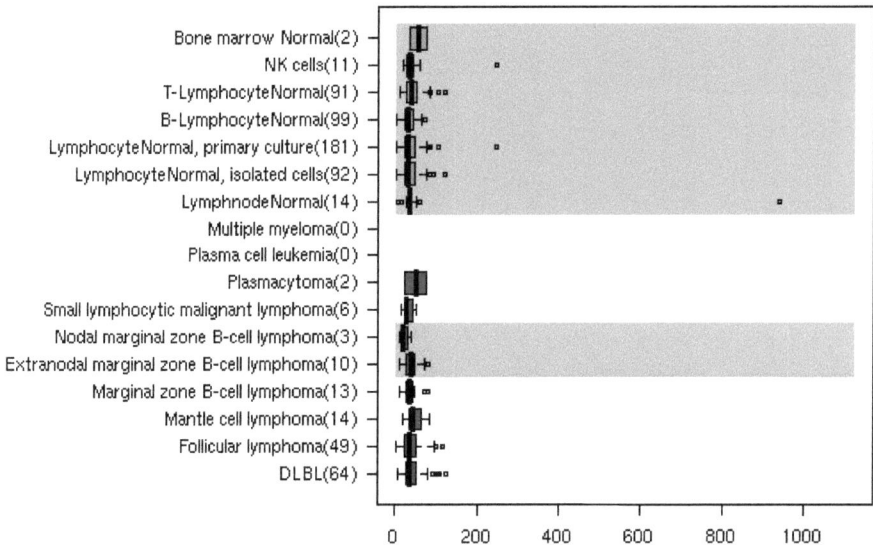

Figure 8. CLCA2 mRNA expression in a panel of lymphomas and B cell leukaemias and their corresponding normal tissues
CLCA2 mRNA expression was analysed using Affymetrix GeneChip technology. The Affymetrix identification code for CLCA2 is 206165_s_at. The expression levels of CLCA2 mRNA are shown in box-plots. The vertical centre line in the box indicates the median. The box itself represents the interquartile range (IQR) between the first and third quartiles. Whiskers extend to 1.5 times the IQR; the position of extreme values, if above the upper whisker limits, are marked by open circles. The human sample collection has been described by the originator of the BioExpress database (GeneLogic Inc.). Numbers in brackets indicate the sample numbers. All samples have been MAS5-normalised. DLBL: Diffuse large B cell lymphoma; other abbreviations see Figure 4.

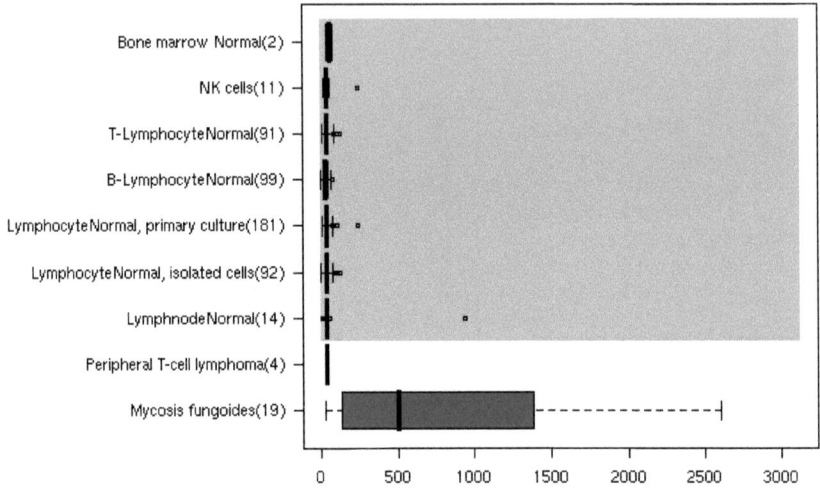

Figure 9. CLCA2 mRNA expression in T cell and B cell lymphomas and their corresponding normal tissues
CLCA2 mRNA expression was analysed using Affymetrix GeneChip technology. The Affymetrix identification code for CLCA2 is 206165_s_at. The expression levels of CLCA2 mRNA are shown in box-plots. The vertical centre line in the box indicates the median. The box itself represents the interquartile range (IQR) between the first and third quartiles. Whiskers extend to 1.5 times the IQR; the position of extreme values, if above the upper whisker limits, are marked by open circles. The human sample collection has been described by the originator of the BioExpress database (GeneLogic Inc.). Numbers in brackets indicate the sample numbers. All samples have been MAS5-normalised. Abbreviations see Figure 4.

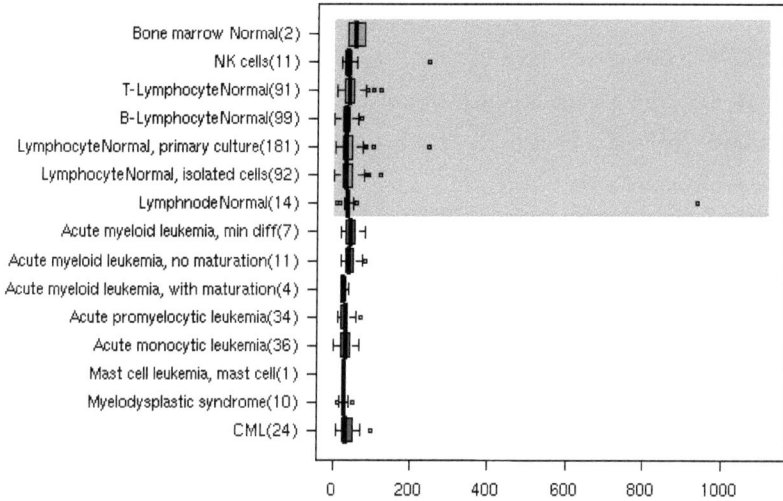

Figure 10. CLCA2 mRNA expression in acute leukaemias such as acute myeloid, acute promyelocytic, and acute monocytic leukeaemia
CLCA2 mRNA expression was analysed using Affymetrix GeneChip technology. The Affymetrix identification code for CLCA2 is 206165_s_at. The expression levels of CLCA2 mRNA are shown in box-plots. The vertical centre line in the box indicates the median. The box itself represents the interquartile range (IQR) between the first and third quartiles. Whiskers extend to 1.5 times the IQR; the position of extreme values, if above the upper whisker limits, are marked by open circles. The human sample collection has been described by the originator of the BioExpress database (GeneLogic Inc.). Numbers in brackets indicate the sample numbers. All samples have been MAS5-normalised. CML: Chronic myelogenous leukemia; other abbreviations sce Figure 4.

RESULTS

As different generations of Affymetrix GeneChips have been used, an *in silico* comparison was performed to test for the robustness and reproducibility of the expression profile of CLCA2. As clearly seen in Figure 11, the expression profile of CLCA2 is independent from the type of GeneChip used. In all cases only SCCs exhibit the dramatic overexpression of CLCA2. For testing the Exon GeneChip generation only a subset of the human lung samples have been used. Although different types of arrays have been used for the expression profiling no statistically relevant variation can be observed in the mRNA levels of the same sample set among different GeneChips. It has to be indicated that for a given tissue sample the relative expression values between the different GeneChips cannot be directly compared with each other (e.g. median for CLCA2 in SCC on U133AB Array is ~ 600, on U133 Plus 2.0 is ~ 800 and on Exon Array ~ 300, respectively).

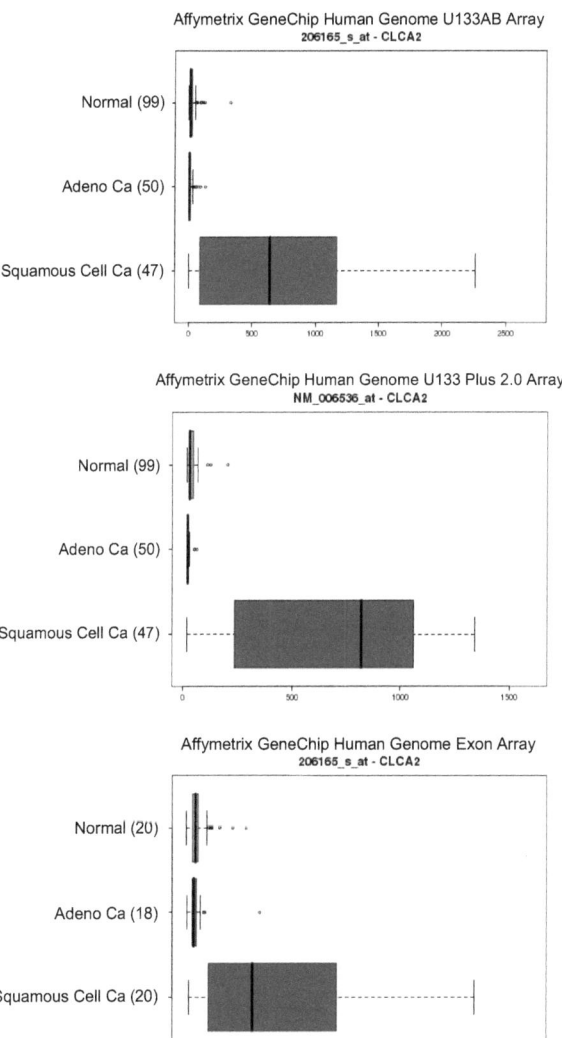

Figure 11. Comparison of CLCA2 expression profiles of a set of human normal and lung tumour (AC and SCC) samples utilising different types of Affymetrix GeneChips
Note that the identification code for CLCA2 is different, depending on the array type used. The expression levels of CLCA2 mRNA are shown in box-plots. The vertical centre line in the box indicates the median. The box itself represents the interquartile range (IQR) between the first and third quartiles. Whiskers extend to 1.5 times the IQR; the position of extreme values, if above the upper whisker limits, are marked by open circles. The human sample collection has been described by the originator of the BioExpress database (GeneLogic Inc.). Numbers in brackets indicate the sample numbers. All samples have been MAS5-normalised. Abbr. see Figure 4.

The infiltration of a number of non-tumourous cell types (e.g. macrophages, lymphocytes) into the tumour tissue and the presence of surrounding non-tumourous tissue (tumour stroma) in whole tissue preparations often leads to falsification in the interpretation of *ex vivo* tissue expression data. To exclude relevant contribution of cells of the tumour stroma, the CLCA2 expression pattern in several laser-capture-microdissected (LCM) tissue samples of lung carcinomas in comparison to non-neoplastic lung tissue samples was analysed (Figure 12). Based on this *in silico* analysis, expression of CLCA2 in tumour stroma could be excluded and the tumour-specific expression pattern of CLCA2 could be confirmed. For a better comparability the total RNA from all cell types and LCM-dissected epithelial cells from different sections of the identical frozen tumour samples was analysed. The expression of CLCA2 transcript is exclusively found in the epithelial tumour cell fraction and if indeed expressed in any stromal compartment, would have led to a decrease in normalised expression levels in epithelial tumour cells, when compared with the corresponding normal tissue. As the relative expression level in the tumour cell population even increased after LCM sample preparation, it can be argued that the detected expression of CLCA2 transcript in all other tissue samples of different origin, which were analysed without LCM dissection, originates exclusively from epithelial tumour cells (see also Figure 2 and Figure 3).

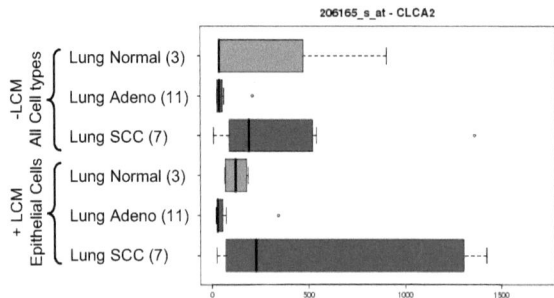

Figure 12. CLCA2 mRNA expression in LCM-dissected lung tumour samples
Samples were anaysed with and without LCM-dissection (with: +LCM / without: -LCM) of epithelial tumour cells. CLCA2 mRNA expression was analysed using Affymetrix GeneChip technology. The Affymetrix identification code for CLCA2 is 206165_s_at. The expression levels of CLCA2 mRNA are shown in box-plots. The vertical centre line in the box indicates the median. The box itself represents the interquartile range (IQR) between the first and third quartiles. Whiskers extend to 1.5 times the IQR; the position of extreme values, if above the upper whisker limits, are marked by open circles. Numbers in brackets indicate the sample numbers. All samples have been MAS5-normalised.

RESULTS

As expression of CLCA2 is detected in esophagus tissue, an interesting aspect was whether CLCA2 is also expressed in other parts of the digestive tract. Therefore the expression levels of CLCA2 were compared in various tissues of the entire digestive tract. As clearly seen in Figure 13, no additional expression is found in other parts of the digestive tract, again underlining the restricted expression pattern of CLCA2.

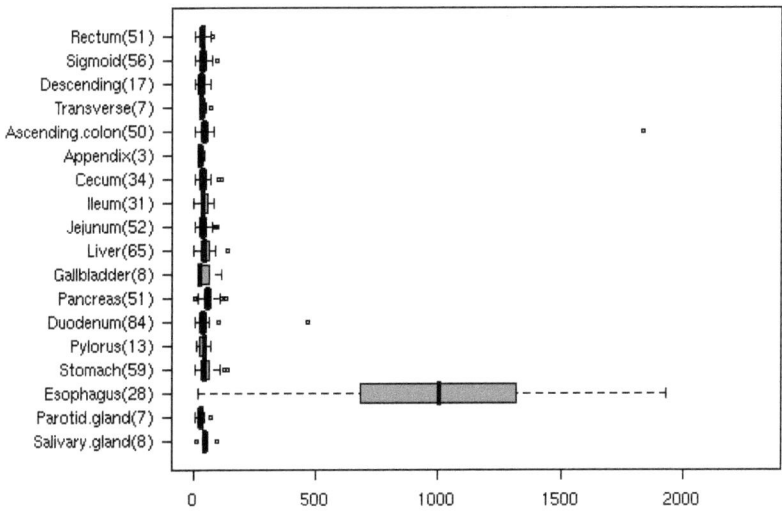

Figure 13. CLCA2 mRNA expression in normal tissues of the digestive tract
CLCA2 mRNA expression was analysed using Affymetrix GeneChip technology. The Affymetrix identification code for CLCA2 is 206165_s_at. The expression levels of CLCA2 mRNA are shown in box-plots. The vertical centre line in the box indicates the median. The box itself represents the interquartile range (IQR) between the first and third quartiles. Whiskers extend to 1.5 times the IQR; the position of extreme values, if above the upper whisker limits, are marked by open circles. The human sample collection has been described by the originator of the BioExpress database (GeneLogic Inc.). Numbers in brackets indicate the sample numbers. All samples have been MAS5-normalised.

RESULTS

That CLCA2 expression might have an influence on development of metastasis as shown in Figure 14. To analyse if there is a correlation of CLCA2 overexpression with development of metastases, it was made use of expression data derived from cell lines and their corresponding developing tumours in xenograft models. It is of great interest that xenografts originating from SSC cell lines such as A431 (cervix carcinoma), H520 (lung carcinoma) and HNOE (head and neck/ esophagus carcinoma) all exhibit a dramatic upregulation of CLCA2 (Figure 14). Except for the pancreatic cell line BxPC3 and MDA-MB-453 (mamma carcinoma) no AC-derived cell line leads to such a pronounced upregulation of CLCA2 in xenografts. In the case of MDA-MB-453 a high expression level of CLCA2 is already observed in 2D culture (Figure 14).

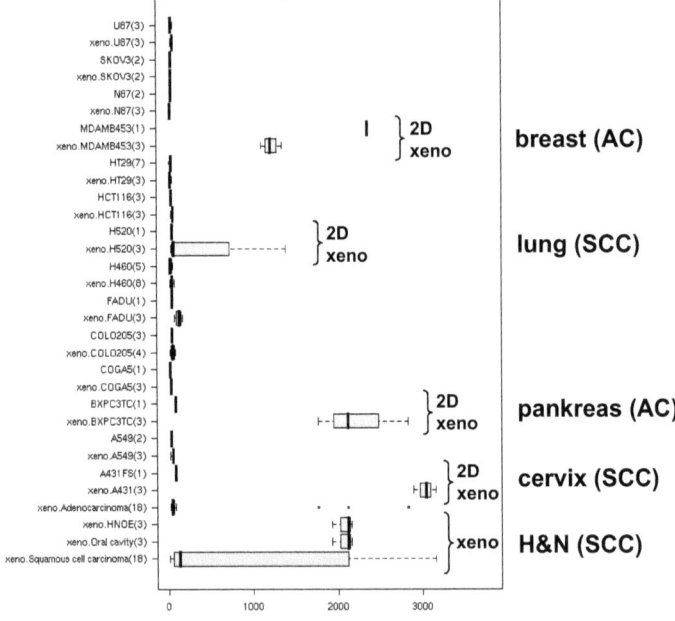

Figure 14. Induction of CLCA2 in xenografts versus 2D cell culture
CLCA2 mRNA expression was analysed using Affymetrix GeneChip technology. The Affymetrix identification code for CLCA2 is 206165_s_at. The expression levels of CLCA2 mRNA are shown in box-plots. The vertical centre line in the box indicates the median. The box itself represents the interquartile range (IQR) between the first and third quartiles. Whiskers extend to 1.5 times the IQR; the position of extreme values, if above the upper whisker limits, are marked by open circles. The human sample collection has been described by the originator of the BioExpress database (GeneLogic Inc.). Numbers in brackets indicate the sample numbers. All samples have been MAS5-normalised. 2D refers to cell lines cultivated in tissue culture and xeno the corresponding xenografts. Other abbreviations see Figure 4.

Expression Profiling of CLCA2 in Tumour Cell Lines

Additional support for a restricted expression profile of CLCA2 comes from the expression profiling analysis of different human carcinoma cell lines (Figure 15). Expression was only detected in the human glioblastoma cell line SNB75, in two SCC cell lines HN5 (head and neck) and A431 (cervix carcinoma) and in the breast carcinoma cell line MDA-MB-453.

Two cell lines were selected for knock-down experiments, HN5 a squamous head and neck carcinoma cell line and MDA-MB-453 (mamma carcinoma) as a representative of an AC cell line with high levels of CLCA2 expression.

RESULTS

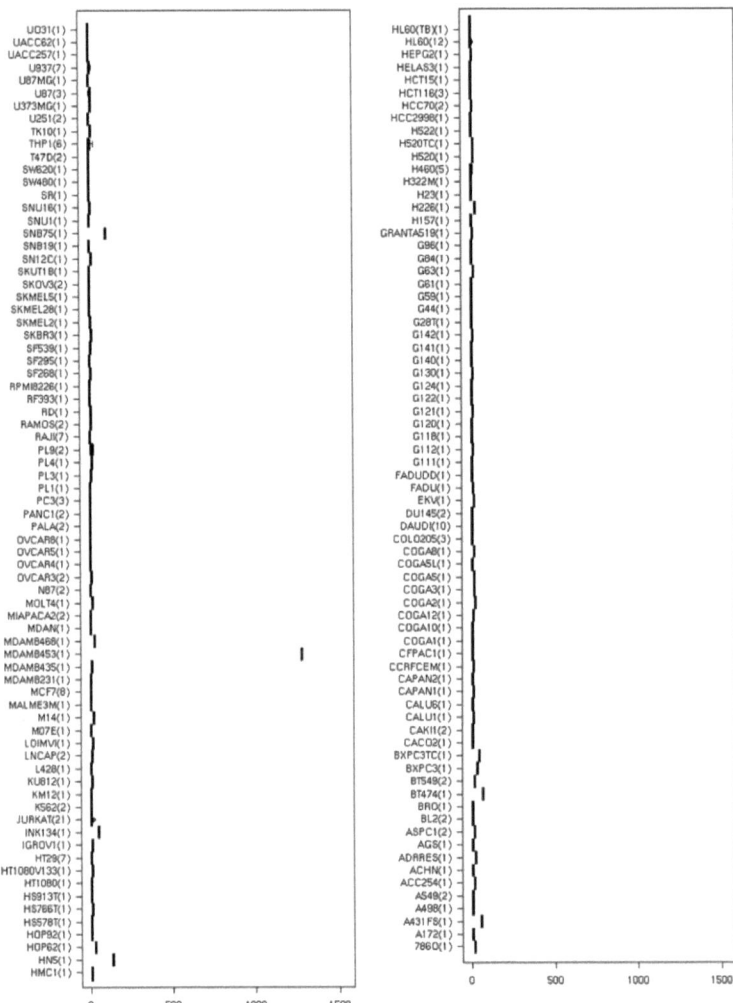

Figure 15. CLCA2 mRNA expression in various human tumour cell lines
CLCA2 mRNA expression was analysed using Affymetrix GeneChip technology. The Affymetrix identification code for CLCA2 is 206165_s_at. The expression levels of CLCA2 mRNA are shown in box-plots. The vertical centre line in the box indicates the median. The box itself represents the interquartile range (IQR) between the first and third quartiles. Whiskers extend to 1.5 times the IQR; the position of extreme values, if above the upper whisker limits, are marked by open circles. Cell line description is according to the nomenclature of the provider (in most cases ATCC (American Type Culture Collection): http://www.atcc.org/home.cfm, DSMZ (Deutsche Sammlung für Mikroorg. und Zellkulturen): http://www.dsmz.de/index.html, and ECACC (European Collection of Animal Cell Cultures): http://fuseiv.star.co.uk/camr/). Numbers in brackets indicate the sample numbers. All samples have been MAS5-normalised. Corresponding tumour types to cell lines are shown in Appendix.

RESULTS

A correlation of CLCA2 expression in metastases of breast cancer with detection of metastases in different sites is shown in Figure 16. To exclude any interference of cells from the tumour stroma, only laser-capture-microdissected (LCM) tissue samples were used for this analysis (for details see Figure 3).

Sample sets which showed no evidence of disease including metastases (NED) were compared to metastases-positive samples (POS). Only samples from patients with a five year observation time have been included in this analysis. Except for metastases to bone, CLCA2 overexpression is obviously needed for formation of metastases in brain, lung, liver, and lymph node respectively. The data shown in Figure 16 strongly supports the results in our xenograft models (Figure 14): CLCA2 expression plays a crucial role in tumour progression and metastasis.

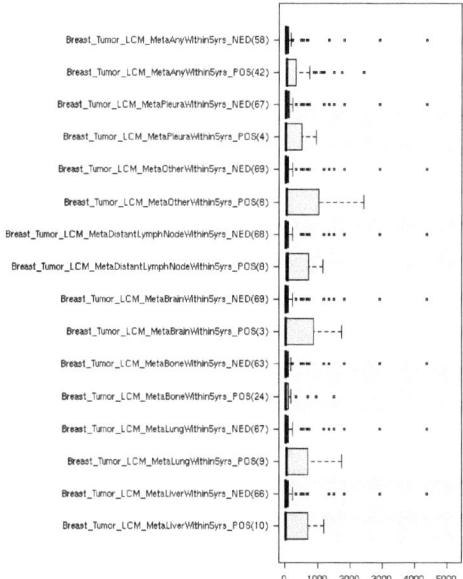

Figure 16. Correlation of metastasis status with CLCA2 expression in LCM breast cancer samples
CLCA2 mRNA expression was analysed using Affymetrix GeneChip technology. The Affymetrix identification code for CLCA2 is 206165_s_at. The expression levels of CLCA2 mRNA are shown in box-plots. The vertical centre line in the box indicates the median. The box itself represents the interquartile range (IQR) between the first and third quartiles. Whiskers extend to 1.5 times the IQR; the position of extreme values, if above the upper whisker limits, are marked by open squares. Numbers in brackets indicate the sample numbers. All samples have been MAS5-normalised. LCM: Laser-capture-microdissected, Meta: Metastases, NED: No evidence of disease (blue), POS: Metastases-positive (yellow). Data generated in the course of the GEN-AU Project (Joint Project: "Genomic Approaches to Tumour Invasion and Metastasis").

A correlation was also found between CLCA2 expression and receptor- and lymphnode status in breast cancer (Figure 17). In the case of Her2 amplification (aggressive status; poor prognosis; [76]) CLCA2 expression is upregulated as compared to tumour samples where Her2 is not amplified. In contrast, if the tumour is still estrogen-sensitive (estrogen receptor status POS; good prognosis; [77]), CLCA2 expression is very low, but becomes upregulated in samples of negative estrogen receptor status (insensitive to estrogen; poor prognosis).

These results support the present view on the essential contribution of CLCA2 to tumour formation, progression, and induction of metastases (Figure 14, Figure 16, Figure 17).

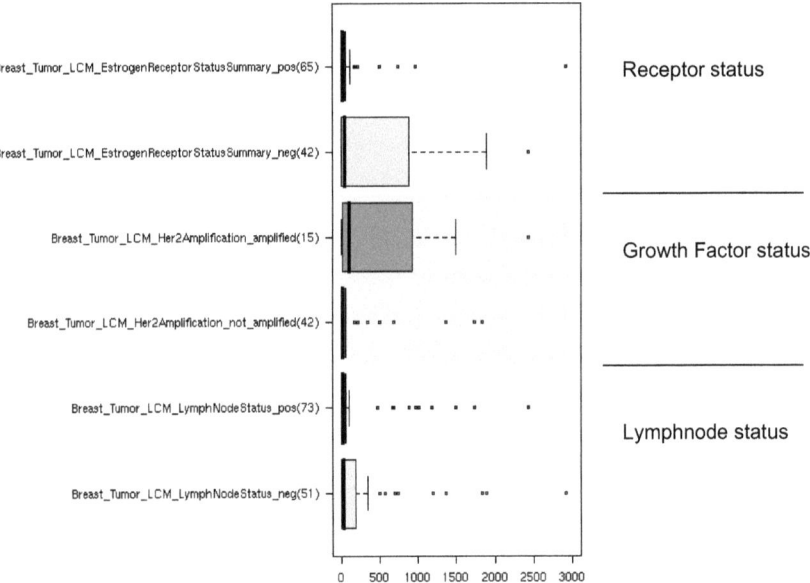

Figure 17. Correlation of receptor and lymphnode status with CLCA2 expression
CLCA2 mRNA expression was analysed using Affymetrix GeneChip technology. The number of samples (n) for each tissue is given. The expression levels of CLCA2 (206165_s_at) mRNA in ER positive and ER negative, in Her2 amplified and Her2 non-amplified and in lymph node positive and negative breast cancer patient samples are shown in box-plots. The vertical centre line in the box indicates the median; the box itself represents the interquartile range (IQR) between the first and third quartiles. Whiskers extend to 1.5 times the IQR. Positions of extreme values are marked by open circles, if they were above the upper whisker limits. Only data from samples which have been laser-capture-microdissected were included in the *in silico* analysis. Data generated in the course of the GEN-AU Project (Joint Project: "Genomic Approaches to Tumour Invasion and Metastasis").

Chromosomal Amplifications of CLCA2 Locus in Tumour Cell Lines

Interestingly, most squamous cell carcinoma cell lines with an increased expression level of CLCA2 also exhibit amplification of the chromosomal locus of CLCA2: 1p22-31 [63], possibly reflecting the link between genomic and transcriptional status in various SCCs. Using Affymetrix SNP-Chip analysis (SNP: Single nucleotide polymorphism), chromosomal amplification or deletion patterns in different human tumours and tumour cell lines were analysed and are shown in Figure 18 and Figure 19. For instance, in FaDu, a head and neck SCC cell line, the copy number (CN) of the CLCA2 locus is amplified (CN ~ 3).

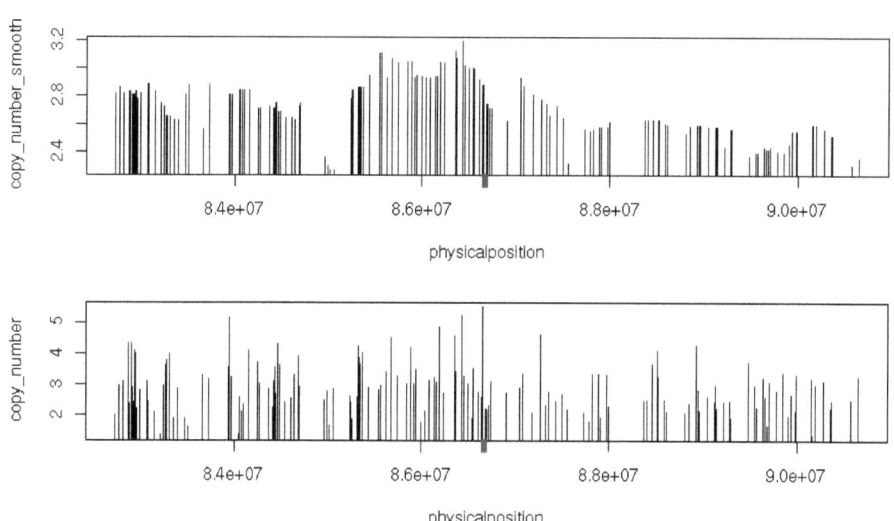

Figure 18. Chromosomal amplification of CLCA2 locus in the SCC cell line FaDu
Using Affymetrix SNP-Chip technology, the chromosomal amplification pattern was analysed in FaDu, a head and neck squamous cell carcinoma cell line. Position of CLCA2 is marked in red. For analysis two algorithms have been used (copy number and copy number smooth).

RESULTS

A set of various colon and breast carcinoma cell lines is shown in Figure 19. Whereas no colon carcinoma cell line contains amplified CLCA2 loci, breast carcinoma cell lines MDA-MB-231, BT-549, and HS-578T exhibit amplification of the CLCA2 loci (marked in yellow).

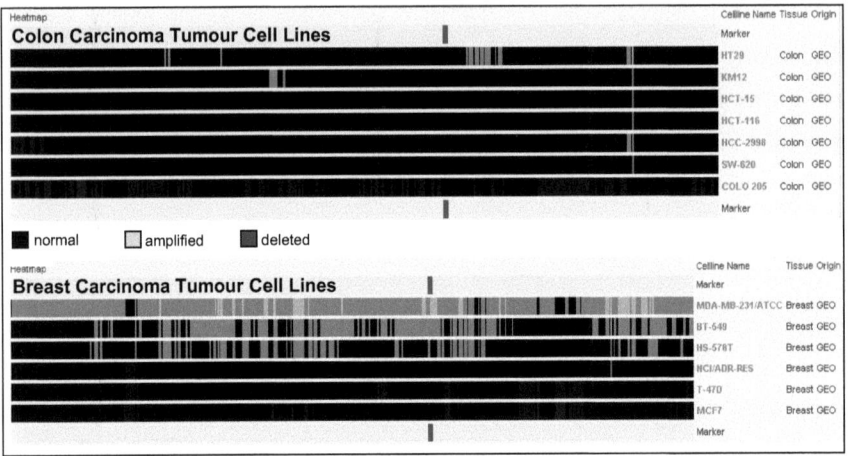

Figure 19. Heatmap of amplification of the chromosomal CLCA2 locus in various human tumour cell lines
Using Affymetrix SNP-Chip technology, the chromosomal amplification pattern was analysed in different human tumour cell lines. Light grey indicated amplified loci, black normal, and dark grey deleted loci. Corresponding tumour types to cell lines are shown in Appendix.

Based on SNP-Chip data and utilising a new algorithm "GLAD" [78] the copy number of genes located at chromome 1 (chr1) has been studied in more detail (higher resolution) in breast cancer patient samples and three SCC cell lines. As shown in Figure 20, not only a small part of the chromosomal region around CLCA2 (1p32-p22) is found to be amplified, but huge parts or even the entire chromosome arm are amplified. In contrast to Her2, no focal amplification pattern was identified.

RESULTS

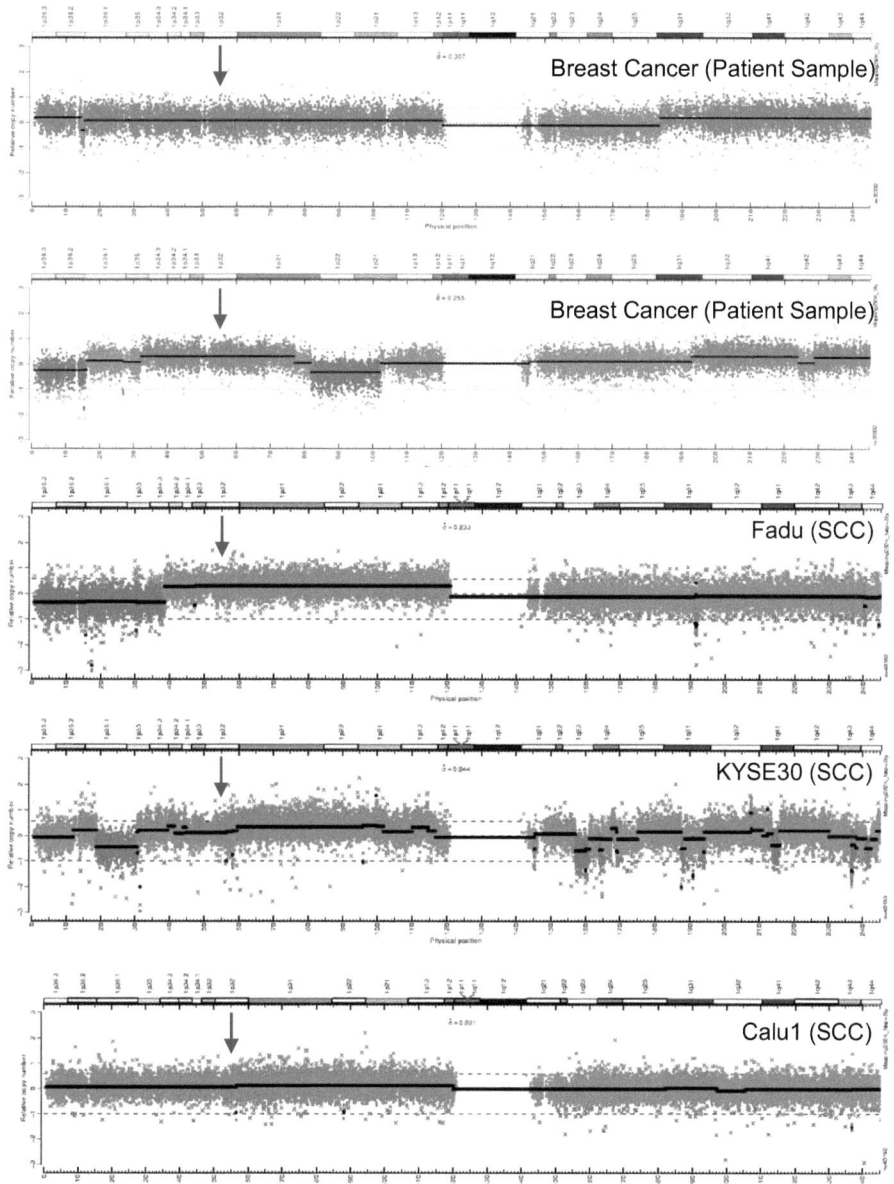

Figure 20. Chromosomal status of chr1 in human breast cancer samples and SCC cell lines
Using Affymetrix SNP-Chip technology, the chromosomal amplification pattern was analysed. Blue indicates amplification, red deletion and green refers to a normal chromosomal status. The arrow indicates the position of CLCA2 at 1p32-p22. Tissue DNA copy numbers were analysed using the GLAD algorithm [78].

Identification of Differential Gene Expression Profiles

To identify transcripts which show a similar expression pattern to that of CLCA2, statistical analyses which correlated normalised expression values of the BioExpress database (GeneLogic Inc.) were performed.

The expression values of CLCA2 were correlated to the expression values of all available probe sets. Thereby the relative differences between the mean expression values of normal and tumourous tissues were used for correlation. In a first analysis the relative expression differences of CLCA2 between normal tissues and SCCs of the lung, cervix, larynx, esophagus, lymphnode metastasis, and skin, respectively, were compared. During the second round of analyses, the focus was laid on the differences in normal tissues and SCCs of the lung, cervix, and lymph nodes.

As expected, all four annotated probe sets for CLCA2 ranked first in both correlation analyses. Among the highest correlated probe sets within both analyses, different markers for stratified epithelia were found. These were, to name but a few, desmocollin (DSC3), desmoglein 3 (DSG3), TP53 apoptosis effector (PERP), plakophilin 1 (PKP1), and keratin 5 (KRT5. The expression of those epithelial differentiation markers seems to be maintained during the transition from a stratified epithelial cell into a hyper-proliferating SCC cell, and could therefore be essential for the establishment of SCCs.

Genes already described to be involved in tumourigenesis, such as ladinin 1 (LAD1/CD18), tripartite motif-containing 29 (TRIM29), stratifin (14-3-3 sigma/SDN), or RAB38, a member of the RAS oncogene family were also identified.

Correlated probe sets which seem to represent yet unidentified markers for stratified epithelia and/or the tumourigenesis of SCCs were found as well.

One of the identified probe sets (219936_s_at), which showed in both analyses nearly an identical expression profile as CLCA2, corresponds to GPR87 and has not been associated with stratified epithelia or the establishment of SCCs before.

The investigation of GPR87 was basis for the PhD-Thesis of Sebastian Glatt with the title "Contribution of hGPR87 and hARHV to squamous cell carcinogenesis", which was done in the Sommergruber lab at Boehringer Ingelheim. The work was recently published in the Int. J. of Cancer [79].

Knock-Down of CLCA2 in CLCA2-Positive Tumour Cells

To minimise off-target effects of the siRNA treatment, commercially available non-targeting control siRNA (siCONTROL#1, Dharmacon; further named "siCTRL") and mock transfections were additionally performed in parallel for all siRNA experiments.

siRNA Experiments in Human Head and Neck SCC Tumour Cell Line HN5

Based on the relatively high expression level of CLCA2 (see section "Expression Profiling"), the cell line HN5 was chosen to study CLCA2 loss-of-function by siRNA-mediated knock-down and its implication on cell survival and proliferation.

Evaluation of CLCA2 siRNAs

First, siGENOME SMART pool (Dharmacon) containing a mixture of four designed siRNAs targeting CLCA2 mRNA, was tested in HN5 cell line for efficient downregulation of the specific mRNA by performing RT-qPCR analysis.

Downregulation of protein levels could not be tested in HN5, because untagged CLCA2 protein could not be detected on Western blots due to lack of an appropriate antibody.

After SMART pool treatment, a reduction of CLCA2 mRNA transcript below 10 % could be determined in HN5 cells (Figure 21).

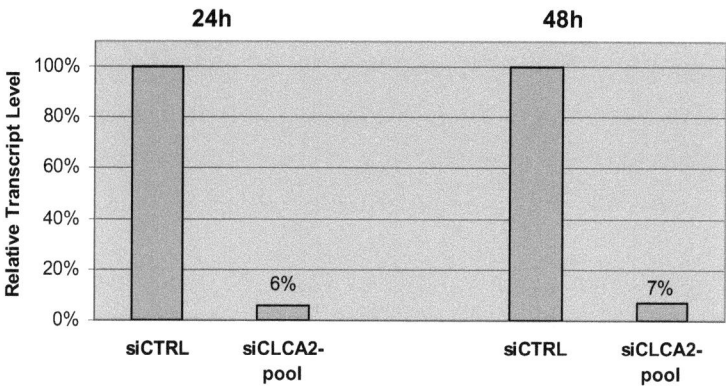

Figure 21. siRNA-mediated knock-down of CLCA2 in human head and neck SCC tumour cell line HN5: Testing siGENOME SMART pool (RT-qPCR)
150,000 cells per well of a 6-well-plate were transfected with CLCA2-siGENOME SMART pool of Dharmacon (siCLCA2-pool), at a concentration of 20 nM siRNA, using Dharmafect 4 transfection reagent. RNA was harvested 24 h and 48 h post transfection. Transcribed cDNA was tested in RT-qPCR using a pool of CLCA2-specific oligonucleotides for the individual siRNAs siCLCA2-2, siCLCA2-4, siCLCA2-5 and siCLCA2-5 – all of Dharmacon; for sequence see section "Material and Methods". CLCA2 mRNA levels were normalised against the endogenous control gene β-2-microglobulin. As control, CLCA2 expression was compared to HN5 cells transfected under same conditions with non-targeting control siRNA CONTROL#1 of Dharmacon (siCTRL).

In a next step, the four individual siRNAs were tested separately under identical conditions and both were shown to exhibit comparable and efficient downregulation of the specific mRNA in RT-qPCR (Figure 22). It was decided, to use only the two most efficient siRNA oligonucleotides, siCLCA2-2 and siCLCA2-4 for further experiments as they lead to a reduction of about 6 % or 10 % at 24 h, and 3 % or 1 % at 48 h, respectively (Figure 22).

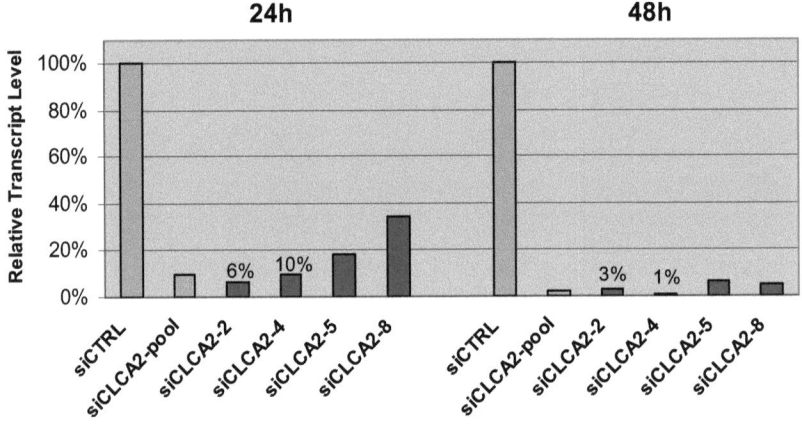

Figure 22. siRNA-mediated knock-down of CLCA2 in human head and neck SCC tumour cell line HN5: Testing individual siRNAs of siGENOME SMART pool (RT-qPCR)
150,000 HN5 cells per well of a 6-well-plate were transfected with CLCA2-siGENOME SMART pool (Dharmacon) (siCLCA2-pool) and in parallel with the individual siRNAs of CLCA2-siGENOME SMART pool (siCLCA2-2, siCLCA2-4, siCLCA2-5, siCLCA2-8) at a concentration of 20 nM siRNA using Dharmafect 4 transfection reagent. RNA was harvested 24 h and 48 h post transfection. cDNA was tested in RT-qPCR using a CLCA2-specific primer pair. CLCA2 mRNA levels were normalised against the endogenous control gene β-2-microglobulin. As control, CLCA2 expression was compared to HN5 cells transfected under identical conditions with non-targeting control siRNA CONTROL#1of Dharmacon (siCTRL).

For optimisation of siRNA concentration, different concentrations of the two most efficient siRNA oligonucleotides were transfected into HN5 cells. In this experiment lysates were analysed after 48 h, because it was shown before, that in general the knock-down effect was more efficient 48 h post transfection (Figure 22).

Furthermore, the two most efficient siRNA oligonucleotides were pooled and analysed for a possible additive effect. As shown in Figure 23, the knock-down efficiency seems to be relatively independent of the siRNA concentration and ranges for all concentrations between 0 and 23 %.

Pooling of the two most efficient siRNA oligonucleotides siCLCA2-2 and siCLCA2-4 seems to slightly improve the knock-down effect (reduction of CLCA2 mRNA transcript to 4-10 %).

In general, efficient knock-down could be achieved under all conditions and higher concentrations did not lead to better results. Therefore, it was decided to perform further experiments at the originally applied concentration of 20 nM siRNA.

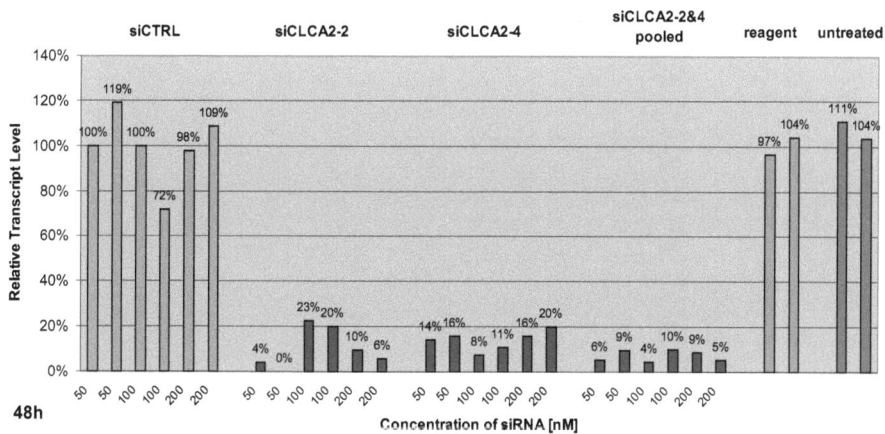

Figure 23. siRNA-mediated knock-down of CLCA2 in human head and neck SCC tumour cell line HN5: Testing different concentrations of siRNA (RT-qPCR)
5,000 HN5 cells per well of a 96-well-plate were transfected with different concentrations of the individual siRNAs siCLCA2-2 and siCLCA2-4 and cells were also transfected with a pool of both (siCLCA2-2&4 pooled) using Dharmafect 4 transfection reagent. RNA was harvested 48 h post transfection. cDNA was tested in RT-qPCR using a CLCA2-specific primer pair. CLCA2 mRNA levels were normalised against the endogenous control gene β-2-microglobulin. As control, CLCA2 expression was compared to HN5 cells transfected under identical conditions with non-targeting control siRNA CONTROL#1of Dharmacon (siCTRL) to untreated HN5 cells (untreated), and to HN5 cells which were treated with transfection reagent alone (reagent).

RESULTS

To analyse if the cell number has an effect on the knock-down efficiency, different cell numbers of HN5 were transfected (6-well-plate) with the two most efficient siRNA oligonucleotides. No significant dependence of the knock-down efficiency on the cell number could be demonstrated (Figure 24). Transfecting either 50,000 or 100,000 cells with siCLCA2-2, lead in both cases to a knock-down of CLCA2 mRNA transcript to 15 %. Using siRNA siCLCA2-4, a knock-down of the CLCA2 transcript to 31 % (50,000 cells) and to 18 % (100,000 cells), respectively, was achieved (Figure 24). Based on these data and previous results the cell number seems not to be the determining factor for knock-down efficacy.

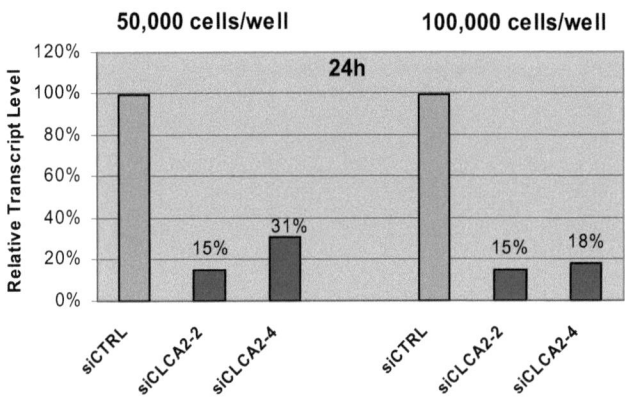

Figure 24. siRNA-mediated knock-down of CLCA2 in human head and neck SCC tumour cell line HN5: Testing different cell numbers (RT-qPCR)
50,000 and 100,000 HN5 cells per well of a 6-well-plate were transfected with the individual siRNAs siCLCA2-2 and siCLCA2-4 at a concentration of 20 nM siRNA using Dharmafect 4 transfection reagent. RNA was harvested 24 h post transfection. cDNA was tested in RT-qPCR using a CLCA2-specific primer pair. CLCA2 mRNA levels were normalised against the endogenous control gene β-2-microglobulin. As control, CLCA2 expression was compared to HN5 cells transfected under identical conditions with non-targeting control siRNA CONTROL#1 of Dharmacon (siCTRL).

Proliferation Assay

To analyse whether knock-down of CLCA2 has any effect on proliferation of HN5 cells, the proliferation of CLCA2 siRNA-treated HN5 cells was analysed in an Alamar Blue Assay. Thereby GPR87 siRNA was used as a positive control. It has been shown previously that downregulation of GPR87 in HN5 cells lead to inhibition of proliferation, to reduction in viability and subsequently to induction of apoptosis [79]. As shown in Figure 25 the IC_{50} values of siCLCA2-2 and of siCLCA2-4 are in the same range of the IC_{50} value for siGPR87, indicating that both, CLCA2 and GPR87 play an essential role for proliferation in HN5 cells.

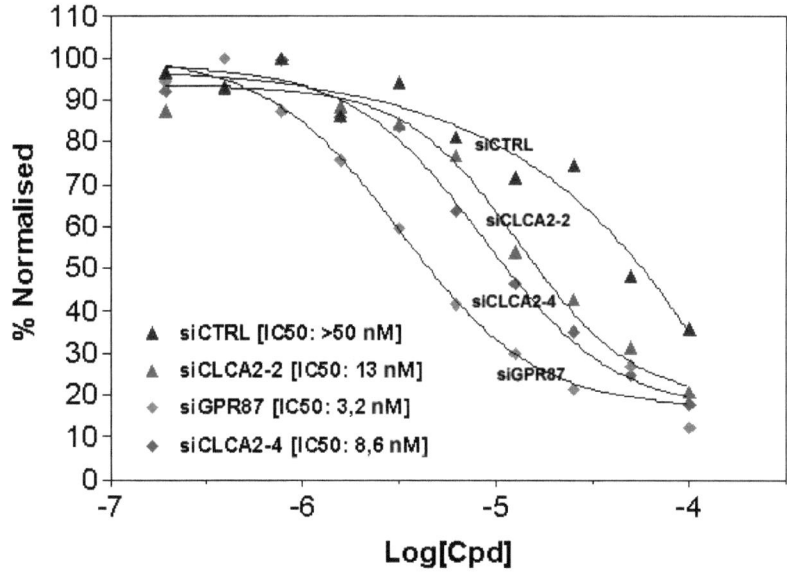

Figure 25. siRNA-mediated knock-down of CLCA2 in human head and neck SCC tumour cell line HN5: Effect on proliferation (Alamar Blue Assay)
5,000 HN5 cells per well of a 96-well-plate were transfected with the individual siRNAs siCLCA2-2 and siCLCA2-4 and as negative control with non-targeting siRNA CONTROL#1 of Dharmacon (siCTRL) and as positive control with GPR87-targeting siRNA siGPR87 at concentrations of 390 pM up to 100 nM siRNA using Dharmafect 4 transfection reagent. For normalisation untreated cells (non-altered proliferation level) and a boiled cell extract (background) was used. Alamar Blue Assay was performed 72 h post transfection.

Apoptosis

Cleavage of PARP (Poly ADP-ribose polymerase), a protein involved in a number of cellular processes which are mainly linked to DNA repair and programmed cell death, is an appropriate indicator of apoptosis as PARP-clevage is observed upon induction of apoptosis. Protein lysates of siRNA-treated HN5 cells were analysed for PARP-cleavage with an anti-PARP-antibody detecting both, full-length and cleaved PARP on Western blot.

48 h after transfection with siCLCA2-4 a pronounced induction of apoptosis (Figure 26) could be detected in HN5 cells, whereas PARP-cleavage in siCLCA2-2-transfected cells was only comparable to the siCTRL-transfected cells. These data again support the previous findings (Figure 25) that siCLCA2-4 is more efficient in knocking-down than siCLCA2-2.

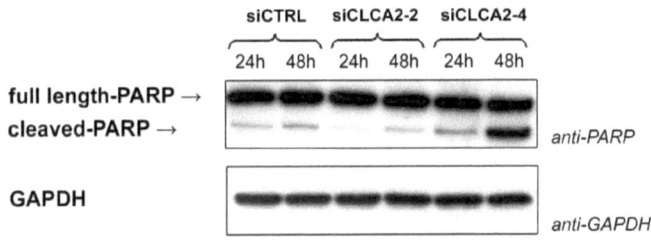

Figure 26. siRNA-mediated knock-down of CLCA2 in human head and neck SCC tumour cell line HN5: Effect on induction of apoptosis (PARP-cleavage)
150,000 HN5 cells per well of a 6-well-plate were transfected with the individual siRNAs siCLCA2-2 and siCLCA2-4 and as control with non-targeting control siRNA CONTROL#1 of Dharmacon (siCTRL) at a concentration of 20 nM siRNA using Dharmafect 4 transfection reagent. Protein lysates were harvested 24 h and 48 h post transfection. For Western blot 50 µg protein was loaded per lane; an anti-PARP antibody was used for detection of cleaved and uncleaved PARP. GAPDH served as loading control.

RESULTS

To not only rely on PARP-cleavage as a read out for induction of apoptosis, a detailed analysis of further effects on changes in apoptotic parameters of siRNA-transfected HN5 cells was performed by investigating the change of nuclear area, nuclear fragmentation, membrane-permeability, mitochondrial status, and nuclear fragmentation. All these processes are linked with induction of apoptosis. For details on the high content screening technology see section "Material and Methods".

In siCLC2-4-transfected HN5 cells dramatical changes in membrane permeability (Figure 27, yellow bar), in the mitochondrial status (Figure 27, black bar) and the nuclear fragmentation (Figure 27, green bar) could be observed, supporting data obtained from PARP-cleavage experiments shown before. As already demonstrated for siCLC2-2-transfected HN5 cells (Figure 25, Figure 26), no significant changes could be detected (Figure 27).

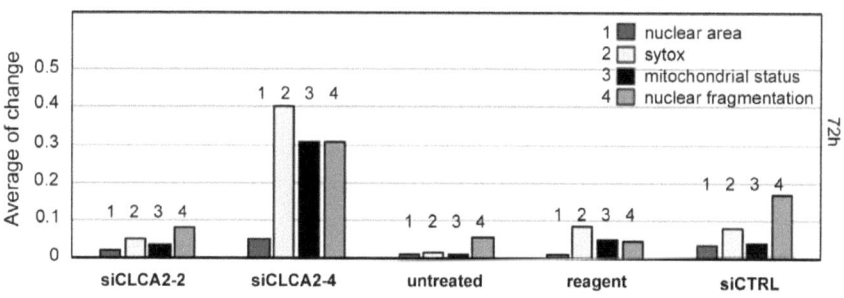

Figure 27. siRNA-mediated knock-down of CLCA2 in human head and neck SCC tumour cell line HN5: Effects on apoptotic parameters (HCS Reader, Cellomics)
5,000 HN5 cells per well of a 96-well-plate were transfected with the individual siRNAs siCLCA2-2 and siCLCA2-4 and as control with non-targeting control siRNA CONTROL#1 of Dharmacon (siCTRL) at a concentration of 20 nM siRNA using Dharmafect 4 transfection reagent. Further controls were untreated cells (untreated) and cells, which were treated with transfection reagent only (reagent). 72 h post transfection, cells were analysed for changes in the apoptotic parameters by using appropriate antibodies for respective stainings which could be evaluated by the ArryScan HCS Reader (Cellomics).

RESULTS

Cell Cycle

The CLCA2 knock-down in siCLA2-2- and siCLCA2-4-treated HN5 cells, leads also to a significant shift in cell cycle 72 h after transfection in comparison to controls (Figure 28). The cell count in G1-phase (2n) was clearly reduced, and cells were arrested in G2, whereby polyploidy was increased. Most likely apoptosis is being initiated in these polyploidic cells (Figure 27). Again, for siCLC2-2-transfected HN5 cells only minor changes of the cell cycle profile could be observed which is in line with previous observations (Figure 25, Figure 26, and Figure 27).

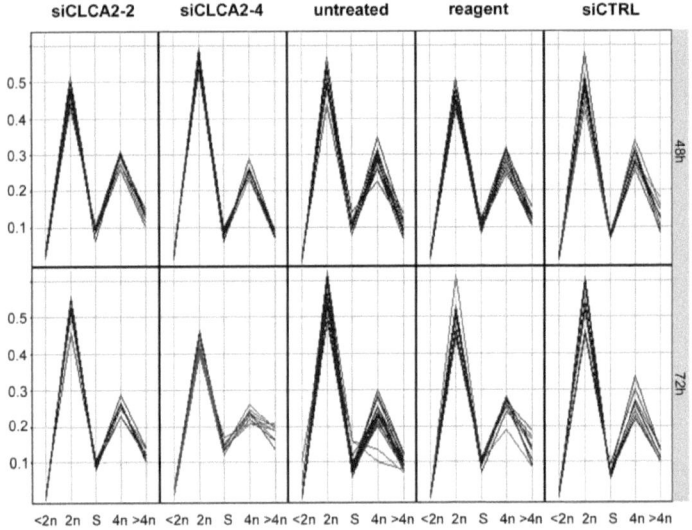

Figure 28. siRNA-mediated knock-down of CLCA2 in human head and neck SCC tumour cell line HN5: Effects on cell cycle (HCS Reader, Cellomics)
5,000 HN5 cells per well of a 96-well-plate were transfected with the individual siRNAs siCLCA2-2 and siCLCA2-4 and as control with non-targeting control siRNA CONTROL#1 of Dharmacon (siCTRL) at a concentration of 20 nM siRNA using Dharmafect 4 transfection reagent. Further controls were untreated cells (untreated) and cells, which were treated with transfection reagent only (reagent). 48 h and 72 h post transfection, cells were stained with DAPI and analysed on the ArryScan HCS Reader (Cellomics).
Red lines indicate statistically relevant changes.
X-axis: The abbreviations <2n, 2n, s, 4n, and >4n represent percentage of cells in sub G1- or G0-phase (<2n), in G1-phase (2n), in S-phase (S), in G2-phase (4n), and polyploidic cells (>4n), respectively.

siRNA Experiments in Human Breast Carcinoma Cell Line MDA-MB-453

According to the BioExpress database (GeneLogic Inc.), the human breast carcinoma cell line MDA-MB-453 exhibits very high CLCA2 expression levels (Figure 15) and was therefore selected for siRNA experiments as well.

As shown in Figure 29, siRNA transfection utilising siCLCA2-2 and siCLCA2-4 revealed that the high levels of naturally expressed CLCA2 in MDA-MB-453 cells could also be knocked-down with these siRNAs to 15 % (siCLCA2-2) and 31 % (siCLCA2-4), respectively, in comparison to non-targeting control siRNA (siCTRL). In contrast to HN5 cells, the efficiency of knock-down is more pronounced with siCLCA2-2 than with siCLCA2-4 (Figure 29, Figure 30). One explanation for this phenomena might be the different transfection reagent (for HN5: Dharmafect 4; for MDA-MB-453: Dharmafect 2) used in this approach (Table 7).

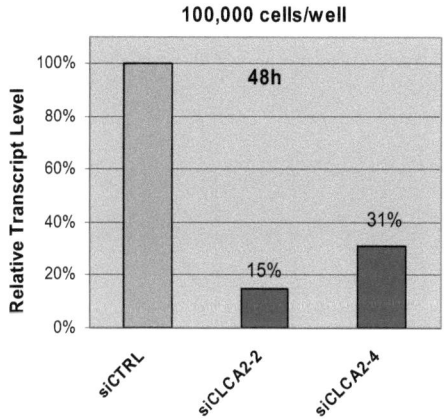

Figure 29. siRNA-mediated knock-down of CLCA2 in breast carcinoma cell line MDA-MB-453: Testing siCLCA2-2 and siCLCA2-4 of siGENOME SMART pool, Dharmacon (RT-qPCR)
100,000 MDA-MB-453 cells per well of a 6-well-plate were transfected with the individual siRNAs siCLCA2-2 and siCLCA2-4, respectively, at a concentration of 20 nM siRNA using Dharmafect 2 transfection reagent. RNA was harvested 48 h post transfection. cDNA was tested in RT-qPCR using a CLCA2-specific primer pair. CLCA2 mRNA levels were normalised against the endogenous control gene β-2-microglobulin. As control, CLCA2 expression is compared to HN5 cells transfected under identical conditions with non-targeting control siRNA CONTROL#1 of Dharmacon (siCTRL).

RESULTS

Next, the siRNA concentration for an efficient downregulation in MDA-MB-453 cells was optimised (Figure 30). Transfection with 50 up to 200 nM siCLCA2-2 results in knock-down of CLCA2 mRNA transcript ranging between 3-12 % compared to cells transfected with non-targeting siRNA (siCTRL). For siCLCA2-4 the results are similar, ranging between 6-12 % but nevertheless siCLCA2-4 seems to be inferior to siCLCA2-2. Co-transfections with both siRNAs are resulting in knock-down of 6-8 %. As the efficiency of downregulation was also demonstrated at lower concentrations of siRNA oligonucleotide and the off-target effects of control siRNA are also less at lower concentrations, it was decided to use a concentration of 20 nM siRNA throughout all further experiments.

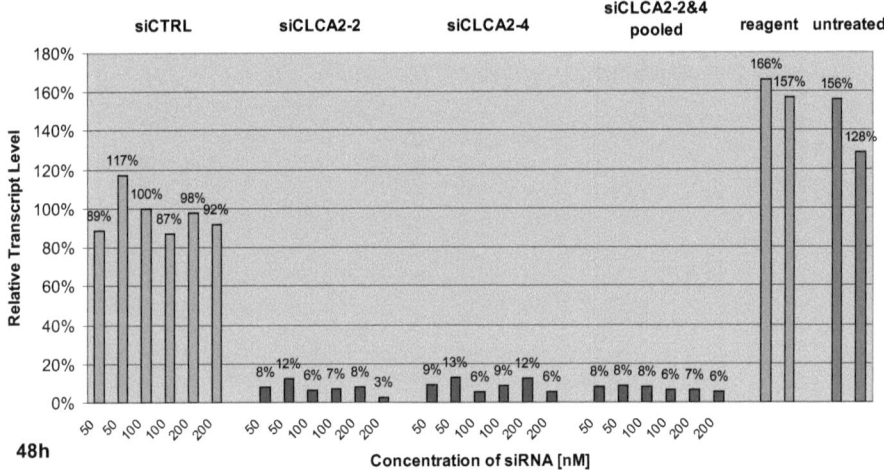

Figure 30. siRNA-mediated knock-down of CLCA2 in human breast carcinoma cell line MDA-MB-453: Testing different concentrations of siRNA (RT-qPCR)
5,000 MDA-MB-453 cells per well of a 96-well-plate were transfected with different concentrations of the individual siRNAs siCLCA2-2 and siCLCA2-4 and cells were transfected with a pool of both as well (siCLCA2-2 and siCLCA2-4 pooled) using Dharmafect 2 transfection reagent. RNA was harvested 48 h post transfection. Transcribed cDNA was tested in RT-qPCR using a CLCA2-specific primer pair. CLCA2 mRNA levels were normalised against the endogenous control gene β-2-microglobulin. As control, CLCA2 expression is compared to MDA-MB-453 cells transfected under same conditions with non-targeting control siRNA CONTROL#1 of Dharmacon (siCTRL), to untreated MDA-MB-453 cells (untreated), and to MDA-MB-453 cells which were treated with transfection reagent only (reagent).

RESULTS

Next, the question whether the cell number per well has an impact on the knock-down efficiency of the CLCA2 mRNA transcript was addressed. As shown in Figure 31, the optimal cell number per well of a 6-well-plate for an efficient knock-down to 10 % is about 50,000. However, increasing the number of cells per well (75,000 to 200,000) leads only to a knock-down efficiency ranging between 26 and 31 %.

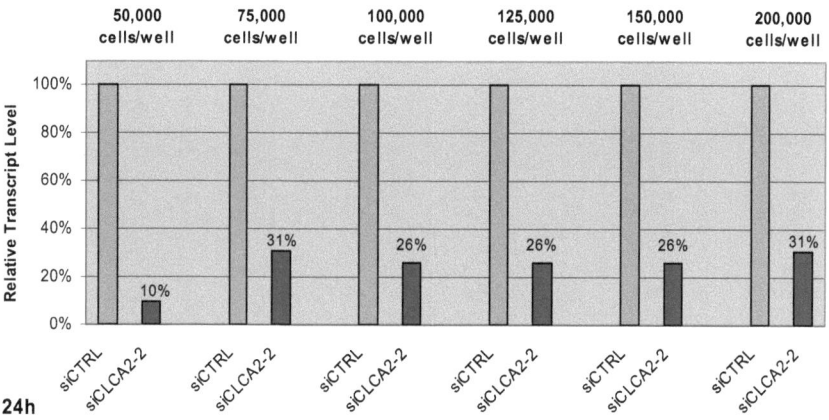

Figure 31. siRNA-mediated knock-down of CLCA2 in human breast carcinoma cell line MDA-MB-453: Testing different cell numbers (RT-qPCR)
As indicated, 50,000 up to 200,000 MDA-MB-453 cells per well of a 6-well-plate were transfected with the individual siRNA (siCLCA2-2) and the control siRNA (siCTRL) each at a concentration of 20 nM using Dharmafect 2 transfection reagent. RNA was harvested 24 h post transfection. cDNA was tested in RT-qPCR using a CLCA2-specific primer pair. CLCA2 mRNA levels were normalised against the endogenous control gene β-2-microglobulin. As control, CLCA2 expression is compared to MDA-MB-453 cells transfected under same conditions with non-targeting control siRNA CONTROL#1 of Dharmacon (siCTRL).

In contrast to HN5 cells, MDA-MB-453 cells were insensitive towards knock-down of CLCA2 when tested in a proliferation assay (Alamar Blue Assay). However, it has to be mentioned that MDA-MB-453 cells only very weakly react with the Alamar Blue reagent and therefore normalisation to untreated MDA-MB-453 cells generated only a small "read out window" (Figure 32).

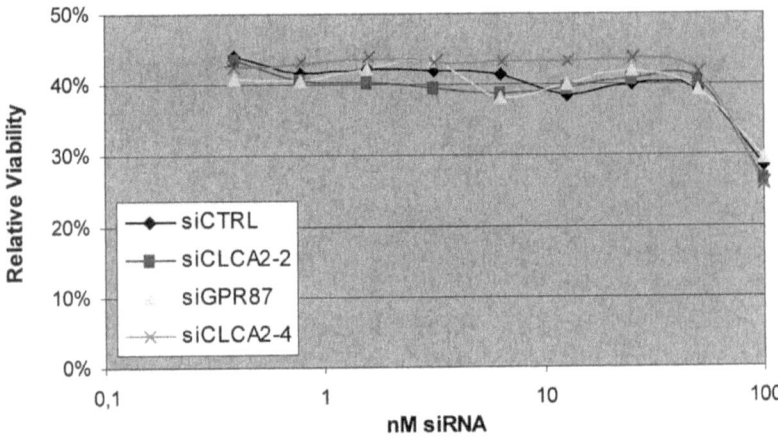

Figure 32. siRNA-mediated knock-down of CLCA2 in human breast cancer cell line MDA-MB-453: Effect on proliferation/viability (Alamar Blue Assay)
5,000 MDA-MB-453 cells per well of a 96-well-plate were transfected with the individual siRNAs siCLCA2-2 and siCLCA2-4 and as negative control with non-targeting siRNA CONTROL#1 of Dharmacon (siCTRL) and as positive control with GPR87-targeting siRNA siGPR87 at concentrations of 390 pM up to 100 nM siRNA using Dharmafect 2 transfection reagent. . For normalisation untreated cells (non-altered proliferation level) and a boiled cell extract (background) was used. Alamar Blue Assay was performed 72 h post transfection.

RESULTS

siRNA Experiments in a Stable CLCA2-Constitutively-Expressing T47D Clone

Due to the lack of a specific antiserum against CLCA2 it was not possible to show siRNA-mediated knock-down on protein level for HN5 or MDA-MB-453 cells.

To proof the efficiency of used CLCA2 siRNAs for knock-down of CLCA2 on the protein level, stable CLCA2-expressing T47D clones were established, with CLCA2 being C-terminally HA-tagged. One of these recombinant cell lines constitutively expressing HA-tagged CLCA2 was used for siRNA transfections: "CLCA2-constitutively-expressing T47D clone #47". For generation of this clone see section "Establishment of Stable wt-CLCA2 Clones".

For Western blot (Figure 33) an anti-HA-antibody detecting both, the full-length CLCA2 and the cleaved C-terminal fragment was used, resulting in two bands on Western blot. In cells transfected with siCLCA2-2 or siCLCA2-4 no or only traces of CLCA2 protein could be detected, whereas in the control cells (untreated cells and cells transfected with non-targeting siRNA) the HA-tagged CLCA2 protein could be visualised as strong bands (Figure 33).

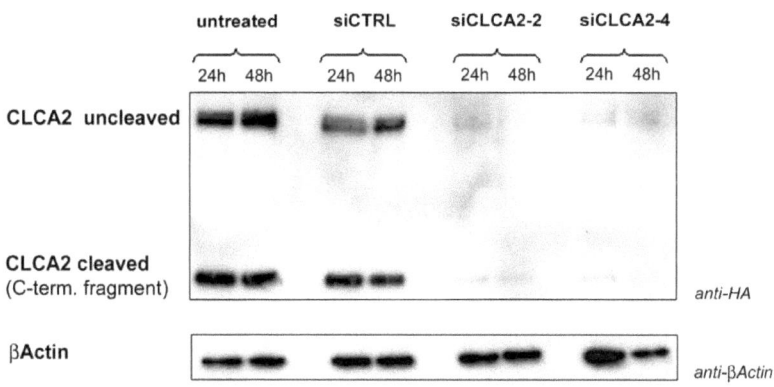

Figure 33. siRNA-mediated knock-down of CLCA2 in CLCA2-constitutively-expressing T47D clone on protein level (Western blot)
150,000 cells of the CLCA2-constitutively-expressing T47D clone #47 per well of a 6-well-plate were transfected with the individual siRNAs siCLCA2-2 and siCLCA2-4 at a concentration of 20 nM siRNA using Dharmafect 4 transfection reagent. As control served untreated cells, and cells transfected with non-targeting siRNA CONTROL#1 of Dharmacon (siCTRL) under same conditions. Protein lysates were harvested 24 h and 48 h post transfection. For Western blot 50 µg protein was loaded per lane; an anti-HA-tag antibody was used for detection of HA-tagged CLCA2. β-Actin served as loading control.

The same protein lysates were further tested for siRNA-mediated PARP-cleavage. In contrast to HN5 cells, no PARP-cleavage could be detected in a CLCA2-constitutively-expressing T47D clone that was treated with CLCA2 siRNA (Figure 34).

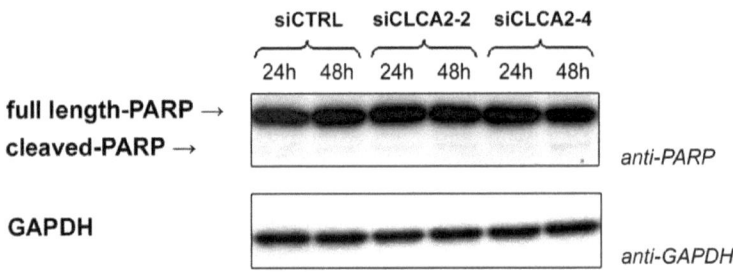

Figure 34. siRNA-mediated knock-down of CLCA2 in CLCA2-constitutively-expressing T47D clone: Effect on apoptosis (PARP-cleavage)
150,000 cells of the CLCA2-constitutively-expressing T47D clone #47 per well of a 6-well-plate were transfected with the individual siRNAs siCLCA2-2 and siCLCA2-4 at a concentration of 20 nM siRNA using Dharmafect 4 transfection reagent. As control non-targeting siRNA CONTROL#1 of Dharmacon (siCTRL) was transfected under same conditions. Protein lysates were harvested 24 h and 48 h post transfection. For Western blot 50 µg protein was loaded per lane; an anti-PARP antibody was used for detection of cleaved and uncleaved PARP. GAPDH served as loading control.

Antibody Generation

Due to the lack of a commercially available CLCA2-specific antibody, it was repeatedly tried to generate specific antisera against human CLCA2. Rabbits were immunised with peptides derived from identified CLCA2-specific epitopes which discriminate between other members of the CLCA family (for details see chapter "Material and Methods").

However, fractions of affinity-purified sera tested on Western blot (data not shown), did not lead to generation of a specific signal. Also peptides linked to proteins such as KHL could not improve the specific immune response. Therefore, it was necessary to generate an HA-tagged CLCA2 expression vector, which – after transfection into the cell line of interest – allowed the detection of CLCA2 on Western blots. However, a polyclonal antiserum was generated which could be used successfully in immunohistochemistry (IHC) studies.

Selection of Peptides for Generation of Antisera

First, specific epitopes which can be used for synthesising small 10 to 20 AA long peptides and for immunisation of rabbits, had to be identified. Appropriate peptides have to exhibit a series of chemical, physical and immunological properties in order to serve as effective immunogens. Algorithms are available which allow the prediction of these characteristics for linear epitopes.

In order to correlate the respective amino acid sequence to secondary structure motifs in CLCA2 (alpha, beta regions as well as turns and coils), mainly the prediction algorithms from Chou and Fasman or from Garnier and colleagues (Figure 35) [80-82] were used. In addition both, hydrophobicity and hydrophilicity play an important role for solubility of the respective epitope peptide, features which essentially contribute to the successful generation of peptide antisera [83-85].

Another important tool for selection of peptide antigens is the determination of the accessability of selected protein regions. Those flexible regions are more likely exposed to the surface of a protein than highly structured regions [86]. The antigenic index prediction of linear/sequential peptide motifs is based on the knowledge of known and experimentally varified peptide epitopes [87]. This approach not only describes the antigenic behaviour of a peptide (protein region) alone, but also takes into account the structural characteristics such as the "protrusion index" [88-92]. BLAST search was performed in order to learn more about the species specificity ("epitope overlapp") of CLCA2; for example, percentage of identity within the CLCA2 peptide epitope among other proteins of the same species (human) is a measure for the background signal to be expected on Western blots. If the epitope is conserved among other species, especially in those animals where the antisera have to be generated (e.g. rabbits), immune tolerance can lead to a weak or almost no immune reaction towards the epitope peptide in the animal. Furthermore, the generation of intra/intra-peptidale secondary structures (e.g. formation of disulfide bridges) was avoided by selecting peptide epitopes without Cys residues. Those secondary structures can efficiently reduce the spatial flexibility of a peptide allowing only a limited number of different configurations in solution and thereby reducing the number of different antibody species in a polyclonal antiserum. Finally, the PROSITE database was used to search for functional epitopes such as Ca-binding motifs, potential glycosylation sites, active sites, etc. within the selected peptide epitopes of CLCA2. For instance, targeting potential glycosylation sites, would lead to less efficient antibodies, as the bulky sugar residues "protect" the immune epitope and by this interfere with the recognition of the linear peptide motif within the CLCA2 [93]. In the absence of any 3D-structure for CLCA2 it could not be speculated, whether the selected peptide epitopes are in principle spatially accessible, as the minimal spatial requirement for an antibody to recognise its epitope is ~ 35 Å. Therefore, it is often observed that high titers towards the peptide can be generated, but the polyclonal antiserum recognises only the protein on IHC slides, but not on Western blots. Based on these algorithms putative peptide epitopes were selected for immunisation, utilising the program "LaserGene 6.1 Protean" comprising almost all algorithms discussed above (Figure 35 and Table 1).

RESULTS

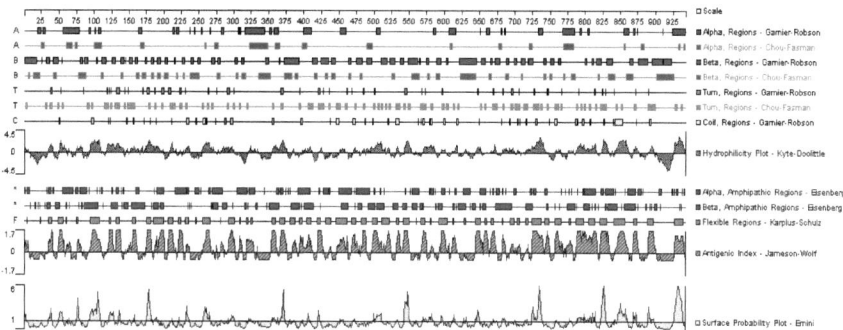

Figure 35. Prediction of structural features of human CLCA2 protein

The indicated programs were used to predict the most promising immunogenic peptide epitopes of CLCA2 for immunising rabbits. The selected peptide epitopes are listed in Table 1.

Amino Acid Pos. in CLCA2 Protein Sequence	Peptide Sequence
K94 – N110	KANNNSKIKQESYEKAN
N179 – D199	NDKPFYINGQNQIKVTRCSSD
E498 – T513	ESTGENVKPHHQLKNT
H690 – H708	HVNHSPSISTPAHSIPGSH
T643 – L656	TVEPETGDPVTLRL

Table 1. Putative immune epitopes selected for synthesis of peptides used to generate antisera

Anti-CLCA2 Polyclonal Peptide Antisera and Their Use for Western Blots

Three rabbits were immunised, two of them with two peptides in parallel – rabbit #1 (K94-N110 + N179-D199), rabbit #2 (E498-T513 + H690-H708) and rabbit #3 (T643-L656). Fractions of affinity-purified sera were tested on Western blot (data not shown), but none of these antibodies generated a specific signal. Also peptides linked to proteins such as KHL could not improve the specific immune response. Therefore, it was necessary to generate an HA-tagged CLCA2 expression vector, which – after transfection into the cell line of interest – allowed a detection of CLCA2 on Western blots.

Immunohistochemistry

In contrast to CLCA2-specific antisera used for Western blotting, one of these affinity-purified antisera exhibits a high titer towards peptide K94-N110 (KANNNSKIKQESYEKAN) and showed a specific staining of tumour cells in human SCC lung tumour samples, utilising a pre-immune serum as a negative control (Figure 36). The different techniques used for antigen retrieval (citrate buffer or proteinase K treatment) lead to slightly different staining pattern; acidic retrieval generate a more spot-like pattern whereas Proteinase K treatment a more diffuse pattern. However, both retrieval methods lead to a specific and pronounced tumour cell staining.

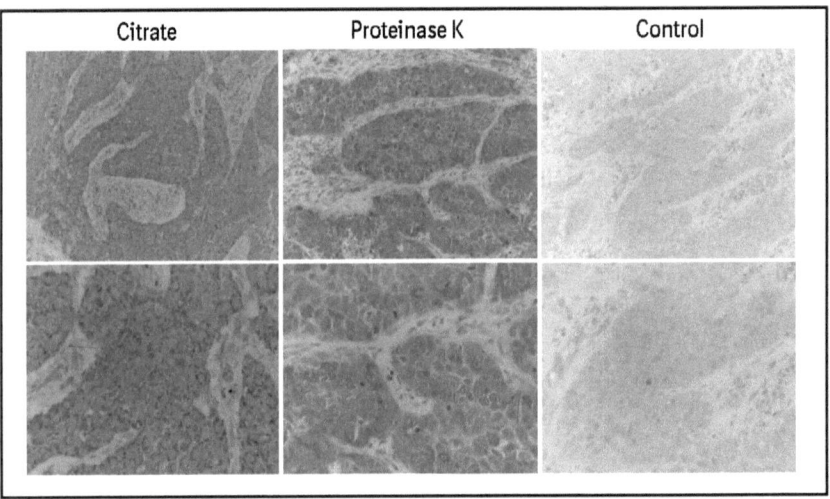

Figure 36. Immunohistochemistry analysis of two human SCC lung cancer samples (NSCLC)
For antigen retrieval samples were either treated with citrate buffer (Citrate: 10 mM, pH: 6.0) or by incubation with Proteinase K (for details see "Material and Methods). As control a pre-immune serum was used for staining generating a weak and unspecific background signal (Control).

Cloning CLCA2

Cloning Various Tags to Expression Vector pCMV-Tag1

pCMV-Tag1 vector (Stratagene), an expression vector with a multiple cloning site and a FLAG-tag downstream was used as a basis vector to introduce other tag sequences than the original FLAG-tag.

Single strand oligonucleotides with the tag sequences of 8F5-tag, HA-tag, V5-tag, myc-tag and His-tag were designed ordered and annealed. Thereby, the oligonucleotides were designed in a way, that each of the annealed fragments bear a BamHI-overhang at the 5'-end and an HindIII-overhang at the 3'-end. Furthermore, the oligonucleotides were designed in a way, that upstream of the HindIII-overhang at the 3'-end, an EcoRV-site was created.

The pCMV-Tag1 vector (Stratagene) was digested with BamHI and HindIII, and the fragment with the respective tag sequence inserted instead of the excised fragment.

After this cloning step a set of pCMV vectors with a multiple cloning site and downstream the tags 8F5, HA, V5, myc, and His were obtained. The created EcoRV-site downstream of the Stop-Codon and downstream of the tag sequence was used for further cloning.

Amplification of CLCA2 and TA-Cloning to pGEM T-Easy Vector

The "CLCA2 Human Full-Length cDNA Clone in pCMV6-XL4" from *Origene* was used as template for amplification.

Full-length CLCA2 was amplified with the primers EBI-15997 and EBI-15958, generating flanking NotI- and BamHI-sites at the ends of the amplified fragment.

This PCR fragment was cloned *via* TA-cloning into the pGEM T-Easy vector. A schematic map of the resulting construct is shown in Figure 37.

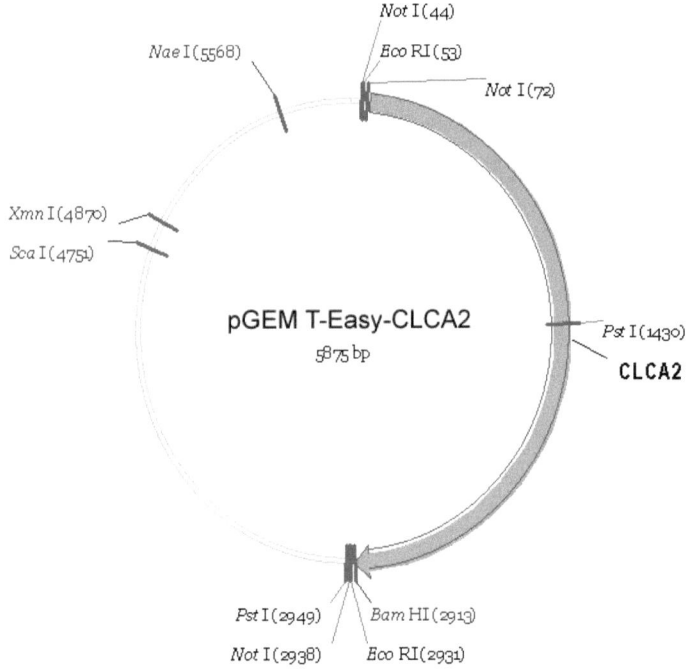

Figure 37. Map of cloning vector pGEM with inserted CLCA2 cDNA
CLCA2 is flanked by a NotI-site (5') and a BamHI-site (3').

RESULTS

The "pGEM T-Easy-CLCA2" vector was digested with BamHI (for cutting out CLCA2) and with ScaI for discriminating with another resulting fragment of nearly the same size. The resulting 1838bp fragment was discarded; the 4037bp fragment was isolated and cut with NotI. The resulting 2841bp fragment was inserted into each of the pCMV-tag vectors (as described above), which had been digested with NotI and BamHI before. As a result, a panel of pCMV expression vectors carrying various tagged CLCA2 genes, were obtained. Figure 38 exemplarily shows the pCMV-CLCA2 vector wherein CLCA2 is tagged with 8F5.

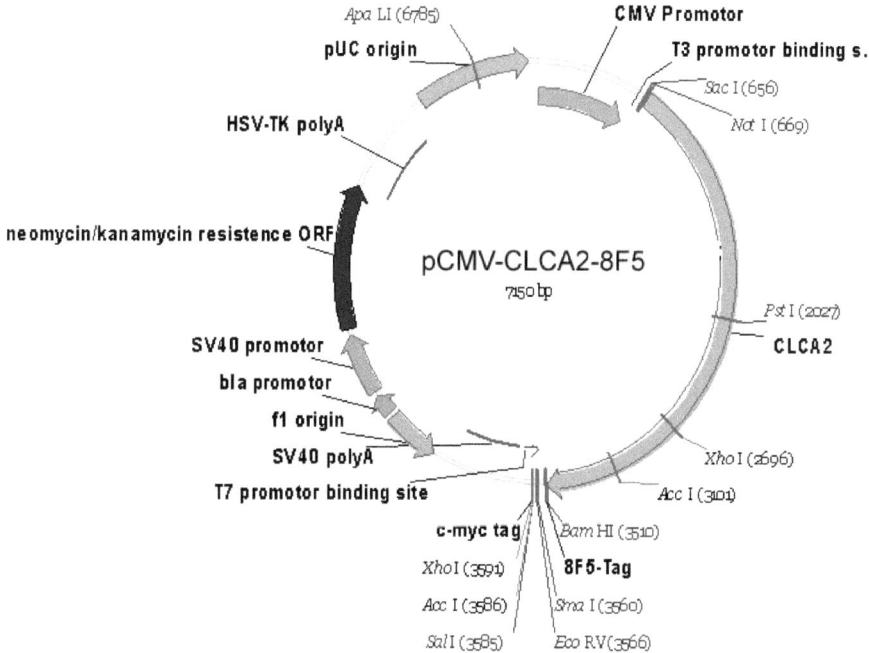

Figure 38. Map of expression vector pCMV-8F5 with inserted CLCA2 cDNA
CLCA2 was inserted between the NotI- and BamHI-site.

Cloning CLCA2-Tag to Expression Vector pTRE2pur

To benefit from an inducible expression system, the differently tagged CLCA2 genes were finally cloned from the pCMV vectors to the pTRE2pur vector of the "Tet-on Gene Expression System" (Clontech) by restriction digestion with NotI and EcoRV, which both are single cutters in the used vectors. Exemplarily, a schematic map of pTRE2pur-CLCA2-HA is shown in Figure 39.

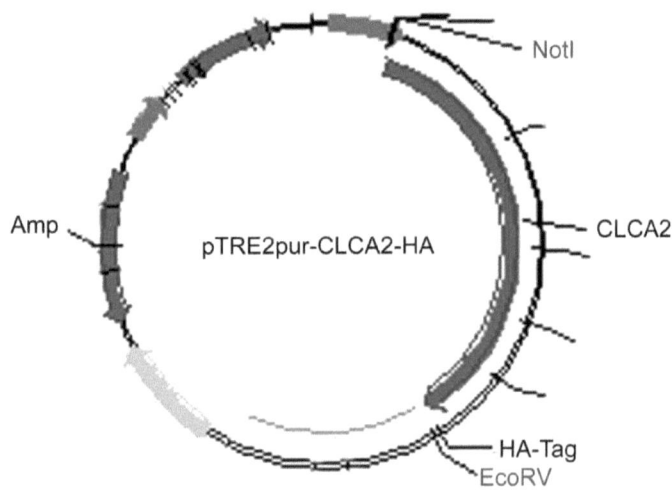

Figure 39. Schematic map of expression vector pTRE2pur with the inserted CLCA2-HA cDNA
The CLCA2-HA-fragment is inserted between the NotI- and EcoRV-site.

Evaluation of Tags on Western Blot

All resulting pTRE2pur-CLCA2-tag expression vectors were individually transfected to T47D Tet-on cells, and lysates of induced and non-induced cells were analysed on Western blot with antibodies detecting the respective tags. As shown in Figure 40, best results were obtained with HA-tag and V5-tag. The cleaved and uncleaved CLCA2-fragment could be detected in the induced sample; the signal in the non-induced sample is extremely low, which demonstrates the strict control of the Tet-on system in T47D Tet-on cells. The strongest signal was found for HA-tagged CLCA2 protein. The antibodies for detection of 8F5 (monoclonal antibody recognising an epitope of VP2 in HRV2 [94]) and myc-tag did not give raise to a valid signal on Western blots and the antibody detecting the His-tag lead to detection of additional bands. Based on these data the construct with the HA-tag was chosen for further investigation.

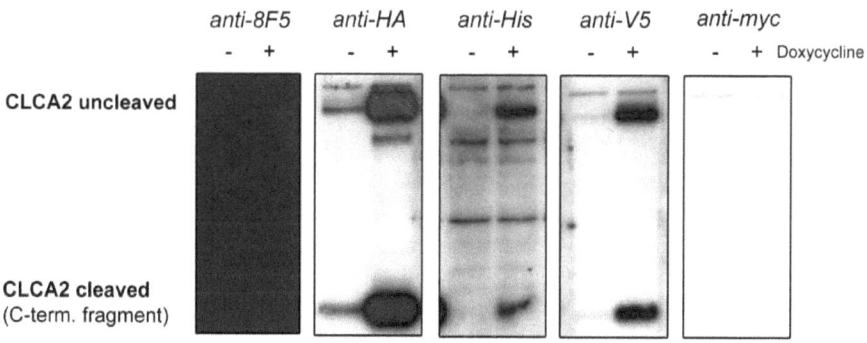

Figure 40. CLCA2 expression with various tags on Western blot
150,000 T47D Tet-on cells were seeded into 6-well-plates and each of pTRE2pur expression vectors containing CLCA2 fused to an 8F5-, HA-, His-, V5-, or myc-tag respectively, was transiently transfected. Expression of tagged CLCA2 was non-induced (-) and induced (+) with Doxycycline 24 h after transfection. Protein lysates were harvested 48 h post induction. For Western blot 50 μg protein was loaded per lane; an antibody detecting the respective tag was used.

Establishment of Stable wt-CLCA2 Clones

As the pTRE2pur-CLCA2-HA construct was chosen for further investigation, this construct was used for establishment of stable wt-CLCA2-expressing clones in the cell line T47D Tet-on, with CLCA2 being C-terminally HA-tagged. The cell line T47D Tet-on was chosen, to benefit from the inducible expression system.

Therefore, 100 clones were picked after selection with Puromycin for 3 weeks and than analysed. All clones were induced and both, induced and non-induced protein lysates were analysed on Western blots with an anti-HA-antibody for detection of HA-tagged CLCA2. The inducible clones were kept in culture and after testing all 100 clones, the most efficiently expressing candidates were compared against each other. In Figure 41 one individual Western blot with five selected inducible clones (#3, 4, 26, 44, 46) is shown, whereby only in induced lysates CLCA2 expression could be detected and non-induced cells did not express detectable amounts of CLCA2.

During the selection process, one clone was identified as a constitutively CLCA2-expressing clone, which wasn't needed to be induced first: "CLCA2-constitutively-expressing T47D clone #47". This clone deemed to be useful to be kept for further experiments and was e.g. used for siRNA experiments (see section "siRNA Experiments in a Stable CLCA2-Constitutively-Expressing T47D Clone").

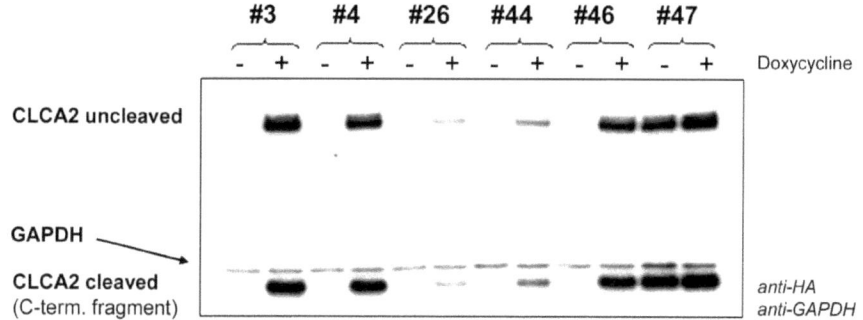

Figure 41. Stable CLCA2-expressing T47D Tet-on clones: Inducible/constitutive expression of HA-tagged CLCA2
Clones #3, #4, #26, #44, #46 expresses CLCA2 inducibly; clone #47 expresses CLCA2 constitutively. Clones were selected for 3 weeks with Puromycin. 300,000 cells were seeded into 6-well-plates and induced with Doxycycline. Protein lysates were harvested 48 h post induction. For Western blot 50 μg protein was loaded per lane; an anti-HA-tag antibody was used for detection of HA-tagged CLCA2. GAPDH served as loading control.

For experiments with an inducible clone, clone #3 ("CLCA2-expressing T47D Tet-on Clone #3") was chosen for further investigations.

To evaluate the concentration of Doxycycline which is necessary for induction of expression of CLCA2, clone #3 was induced with various concentrations and CLCA2 expression was analysed on Western blot. As shown in Figure 42 the expression of CLCA2 could not be dramatically enhanced, using more than 2 µg/ml Doxycycline in media. So this concentration was chosen for all further experiments.

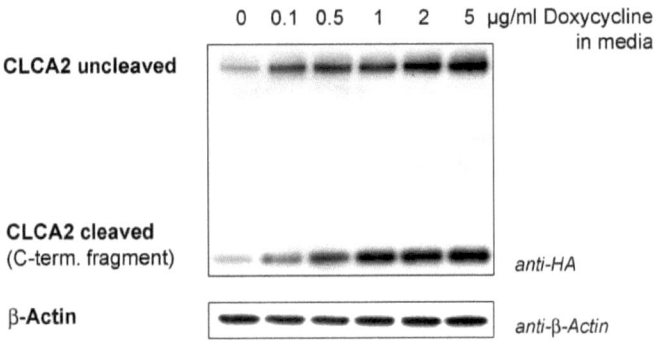

Figure 42. Doxycline-induced induction of CLCA2 expression in stable T47D Tet-on clone: Concentration-dependency
300,000 cells were seeded into 6-well-plates and CLCA2 expression induced with 0, 0.1, 0.5, 1, 2, 5 µg/ml Doxycycline in media. Protein lysates were harvested 48 h post induction. For Western blot 50 µg protein was loaded per lane; an anti-HA-tag antibody was used for detection of HA-tagged CLCA2. β-Actin served as loading control.

In Figure 43 a time-kinetic of expression of CLCA2 in clone #3 is shown. In parallel, cells were induced and non-induced and protein lysates harvested after variable time-periods. CLCA2 expression was analysed on Western blots.

CLCA2 expression could be detected already 24 h post induction, after 48 h expression was not further increased.

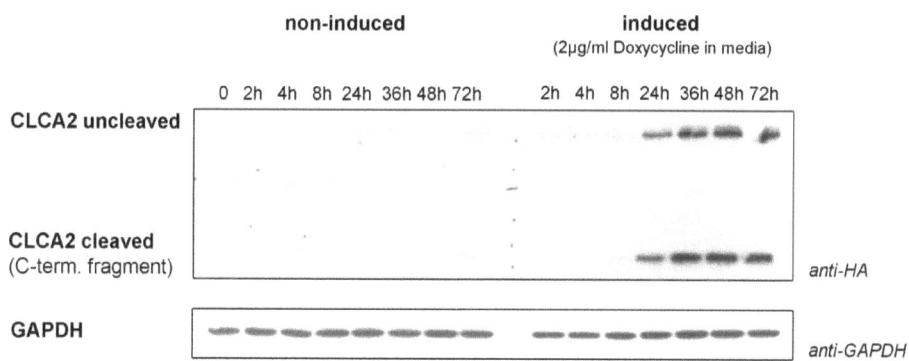

Figure 43. Doxycline-induced induction of CLCA2 expression in stable T47D Tet-on clone: Time-dependency
300,000 cells were seeded into 6-well-plates and non-induced/induced with 2 µg/ml Doxycycline in media. Protein lysates were harvested 2 h, 4 h, 8 h, 24 h, 36 h, 48 h and 72 h post induction. For Western blot 50 µg protein was loaded per lane; an anti-HA-tag antibody was used for detection of HA-tagged CLCA2. GAPDH served as loading control.

These results clearly demonstrate the time-dependent induction of CLCA2 expression, which was only in induced lysates of clone #3 detectable, but no in protein lysates of non-induced cells.

Proteolytic Processing

In May 2006 Pawlowski published that CLCA1 has a hydrolase domain responsible for processing mature full-length CLCA1 [65]. As from peptide-design for immunisation (see above) it was already known, that the homology within the CLCA family is quite high, it was assumed that CLCA2 might also contain such a hydrolase domain. Therefore, experiments with matrixmetalloprotease-inhibitors such as Phenantrolin, Diprotin B, Bestatin, Amastatin and others were performed as the internal hydrolase activity of CLCA1 was supposed to be a member of the metalloproteinase family (data not shown). Besides Phenantrolin, no effect on cleavage of CLCA2 by treatment with these inhibitors (at concentrations recommended by conventional literature) could be demonstrated. As Phenantrolin was quite toxic for the CLCA2-expressing T47D Tet-on cell line, no final conclusion could be drawn from this experiment. To learn more about the putative internal hydrolase activity of CLCA2, first an alignment of CLCA1 and CLCA2 protein sequence was made (Figure 45) with the idea that the hydrolase domain which was found by Pawlowski in CLCA1 might be conserved in the whole protein family. Indeed, a high homology within the hydrolase domain of CLCA1 and respective segment in CLCA2 (equivalent sequence in CLCA2 is located at AA 29-208) could be identified (Figure 44).

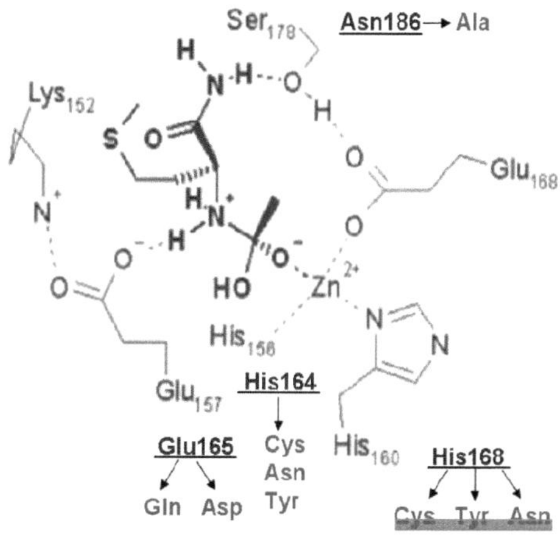

RESULTS

Figure 44. Catalytic unit of hydrolase domain of CLCA1 and corresponding sites in CLCA2
Schematic representation showing polar interactions (dashed lines) between sTI (bold) and putative catalytic site residues.
Sites within the illustration indicate the respective AA positions of CLCA1, underlined sites refer to corresponding positions in CLCA2 and letters followed by an arrow describe planed mutations within the catalytic center of the putative hydrolase domain of CLCA2.

Figure 45. Amino acid sequence alignment of CLCA1 and CLCA2
Protein Sequences of CLCA1 and CLCA2 were aligned utilising CLC Bio.

As Cysteines not only play an essential role in the formation of protein structure, but also can be part of the catalytic site, those Cys residues within the proposed hydrolase domain were identified, which are conserved between CLCA1 and CLCA2. These Cysteines are in CLCA2 protein sequence located at the following amino acid (AA) positions: C132, C196, C206, C211, and C216 (Figure 46). These Cys residues were also selected for mutational analysis.

Mutating Conserved Cysteines

The conserved Cysteines within CLCA1 and CLCA2 protein structure (in CLCA2: C132, C196, C206, C211, and C216) were mutated to Alanine and in the case of C132 to Alanine and Serine. These point mutations were directly introduced into the expression vector pTRE2pur-CLCA2-HA carrying the HA-tagged wt-CLCA2, by utilisation of the "QuikChange II XL Site-Directed Mutagenesis Kit" (Stratagene). Mutation constructs were sequenced and transiently transfected into T47D Tet-on cells, where the expression of mutated CLCA2 was induced. Protein lysates were tested on Western blot using an anti-HA-antibody.

As shown in Figure 46, cleavage of CLCA2-bearing mutations C211A or C216A is heavily impaired, whereas mutations at C132, C196, and C206 completely abolish the proteolytic processing.

RESULTS

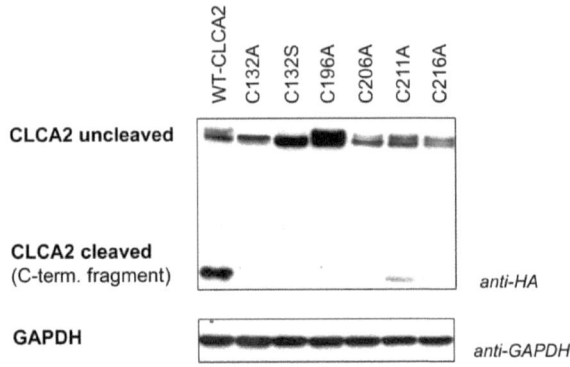

Figure 46. Mutations of conserved Cysteines of CLCA2 and their effects on its proteolytic processing
150,000 T47D Tet-on cells were seeded into 6-well-plates and transiently transfected with various constructs of pTRE2pur expression vector, expressing either HA-tagged wt-CLCA2 or bearing one of the following mutations: C132A, C132S, C196A, C206A, C211A, and C216A. Expression was induced with Doxycycline 24 h post transfection. Protein lysates were harvested 48 h post induction. For Western blot 50 µg protein was loaded per lane; an anti-HA-tag antibody was used for detection of HA-tagged wt/mutant CLCA2. GAPDH served as loading control.

Mutational Analysis of the Putative Hydrolase Domain

As the homology search with CLCA1 and CLCA2 protein sequences predicted a putative hydrolase domain for CLCA2 for AA 29-208, it was decided to point mutate respective sites in CLCA2 which might be involved in the formation of the catalytic site. Figure 47 summarises the influence of the mutations within the putative hydrolase domain on the proteolytic cleavage of CLCA2. N186A is the only mutation being tolerated, all other mutations completely abolish cleavage of CLCA2.

RESULTS

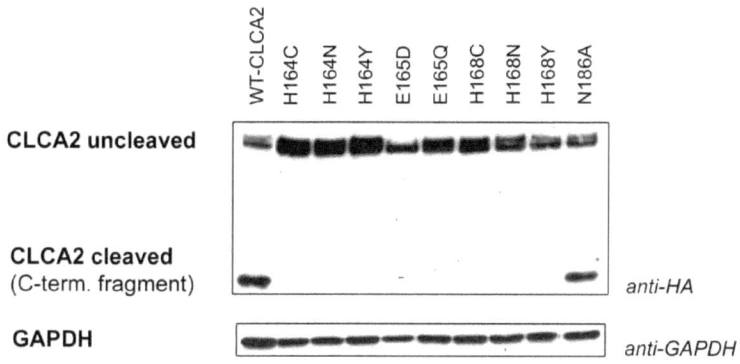

Figure 47. Mutations in the putative hydrolase domain of CLCA2 and their effects on its proteolytic processing
150,000 T47D Tet-on cells were seeded into 6-well-plates and transiently transfected with various constructs of pTRE2pur expression vector, expressing either HA-tagged wt-CLCA2 or bearing one of the following mutations: H164C, H164N, H164Y, E165D, E165Q, H168C, H168N, H168Y, and N186A. Expression was induced with Doxycycline 24 h post transfection. Protein lysates were harvested 48 h post induction. For Western blot 50 µg protein was loaded per lane; an anti-HA-tag antibody was used for detection of HA-tagged wt/mutant CLCA2. GAPDH served as loading control.

In Figure 48 a schematic overview of the impact on the proteolytic processing of all analysed mutations within the putative hydrolase domain of CLCA2 is presented.

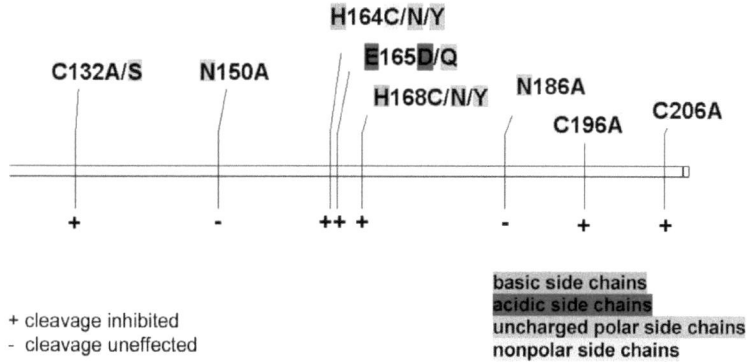

Figure 48. Summary of effects of mutations within the putative hydrolase domain of CLCA2
+ indicates complete inhibition of CLCA2 cleavage as analysed on Western blots (no cleaved fragment detectable)
- indicates no influence of the mutation on CLCA2 cleavage (Western blot pattern similar to that of wt-CLCA2)
E (Glutamic Acid) and D (Aspartic Acid) are amino acids with an acidic side chain; H (Histidine) has a basic side chain; C (Cysteine) and A (Alanine) have a non-polar side chain; S (Serine), N (Asparagine), Y (Tyrosine), Q (Glutamine) have a polar side chain.

Mutational Analysis of the Putative Cleavage Site

As known from bioinformatic analysis and literature [95], there is a mono basic cleavage site predicted at position R674.

Mutational analyses were first done by Alanine-walking starting from predicted positions P5 to P4' (R670 to R678) by subsequent and stepwise substitution of the respective amino acid to an Alanine. These studies were further improved by introducing (especially at position R674 of the predicted mono basic cleavage site) at least one of each class of amino acids (i.e. amino acids with either a basic, acidic, non-polar or uncharged polar side chain). Western blot data on cleavage efficiency of mutated CLCA2 proteins are shown in Figure 49 and the effects are summarised in Figure 50. These data indicate that position R674 might serve as the P1 cleavage site, as this position strictly not allows any exchange of Arginine, besides Lysine which is also tolerated, due to the fact, that both are amino acids with basic side chains.

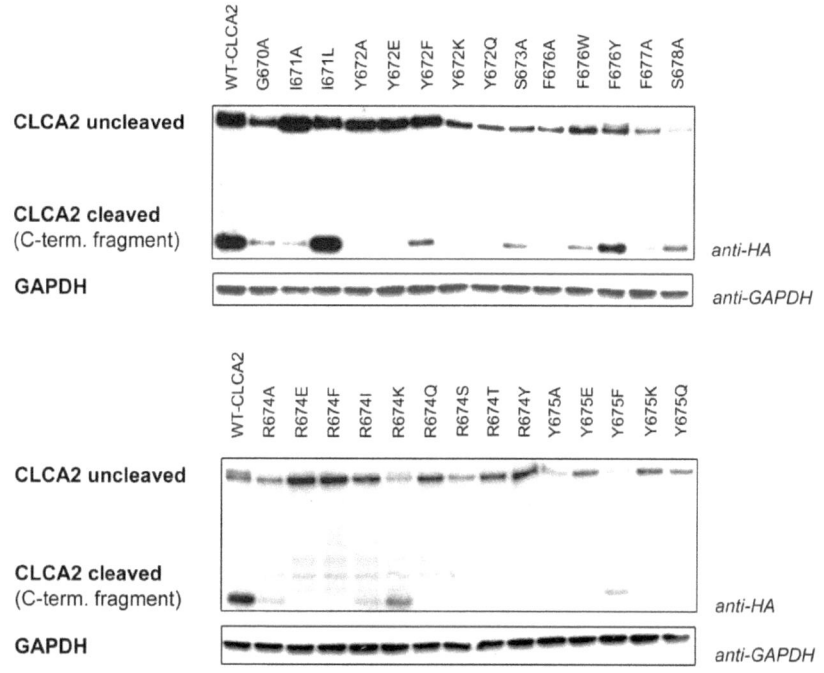

Figure 49. Mutations in the putative cleavage site of CLCA2 and their effects on its proteolytic processing
150,000 T47D Tet-on cells were seeded into 6-well-plates and transiently transfected with various constructs of pTRE2pur expression vector, expressing either HA-tagged wt-CLCA2 or bearing one of the following mutations: 670A, I671A/L, Y672A/E/F/K/Q, S673A, F676A/W/Y, F677A, F678A, R674A/E/F/I/K/Q/S/T/Y and Y675A/E/F/K/Q. Expression was induced with Doxycycline 24 h post transfection. Protein lysates were harvested 48 h post induction. For Western blot 50 µg protein was loaded per lane; an anti-HA-tag antibody was used for detection of HA-tagged wt/mutant CLCA2. GAPDH served as loading control.

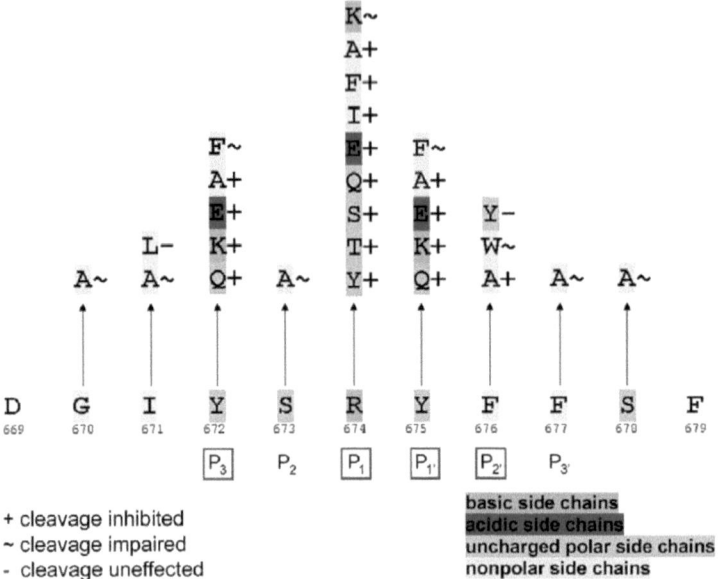

Figure 50. Summary of effects of mutations in the putative cleavage site of CLCA2 on its proteolytic processing
+ indicates complete inhibition of CLCA2 cleavage by analysis on Western blot (no cleaved fragment detectable)
~ indicates impaired CLCA2 cleavage (detectable cleavage fragment on Western blot is reduced in comparison to wt-CLCA2)
- indicates no influence on CLCA2 cleavage (pattern on Western blot is similar with that of wt-CLCA2)
E (Glutamic Acid) and D (Aspartic Acid) are amino acids with an acidic side chain; H (Histidine) has a basic side chain; C (Cysteine) and A (Alanine) have a non-polar side chain; S (Serine), N (Asparagine), Y (Tyrosine), Q (Glutamine) have a polar side chain.
Boxed positions indicate most sensitive residues within predicted cleavage site.

Establishment of Stable Mutant-CLCA2 Clones

As for wt-CLCA2, it was decided to establish stable cell lines, expressing mutated CLCA2, which are devoid of cleavage ability. Therefore, constructs with either point mutations in the catalytic unit (E165Q or H186N), or with a point mutation in the monobasic cleavage site were transfected into T47D Tet-on cells.

All mutations were originally directly introduced into the expression vector pTRE2pur-CLCA2-HA carrying the HA-tagged wt-CLCA2, by utilisation of the "QuikChange II XL Site-Directed Mutagenesis Kit" (Stratagene). The cell line T47D

Tet-on was chosen, to benefit from the inducible expression system and to allow comparison to wt-CLCA2 clones.

For establishment of stable clones, 100 clones were picked after selection with Puromycin for 3 weeks and analysed. All clones were induced and protein lysates (induced and non-induced) were analysed on Western blots utilising an anti-HA-antibody for detection of (wild type/mutated) CLCA2. Inducible clones were kept in culture and after testing 100 clones, the most promising clones were compared against each other. In Figure 51 Western blot analysis of the most efficiently inducible clones for various CLCA2 mutants is shown.

Double bands seen for HA-tagged CLCA2 on Western blots for e.g. clones #3 and #32 might be due to incomplete glycosylation as a result of tremendous overexpression.

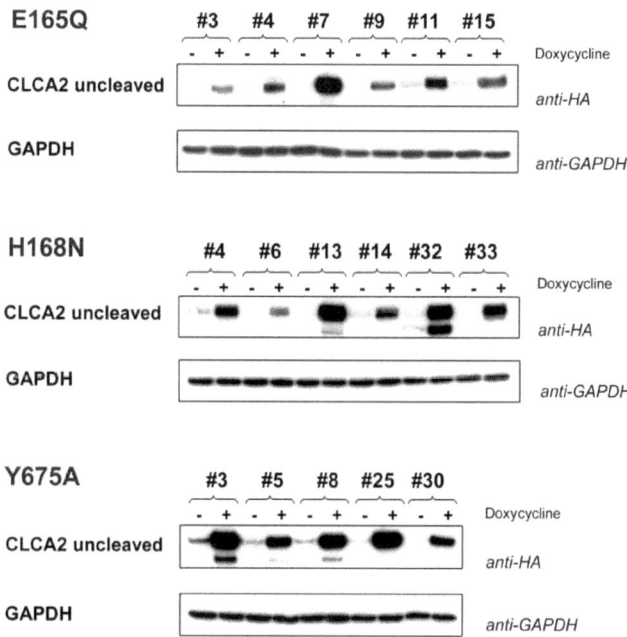

Figure 51. Stable mutant-CLCA2-expressing T47D Tet-on clones:
Inducible/constitutive expression of HA-tagged CLCA2 mutants E165Q, H168N, Y675A
Clones were selected for 3 weeks with Puromycin. 300,000 cells were seeded into 6-well-plates and induced with Doxycycline. Protein lysates were harvested 48 h post induction. For Western blot 50 µg protein was loaded per lane; an anti-HA-tag antibody was used for detection of HA-tagged wt/mutant CLCA2. GAPDH served as loading control.

RESULTS

The following clones were chosen for further investigations:
- mutant-CLCA2(E165Q)-expressing T47D Tet-on clone #7
- mutant-CLCA2(H168N)-expressing T47D Tet-on clone #13
- mutant-CLCA2(Y675A)-expressing T47D Tet-on clone #25

Inhibition of Post-Translational Modifications

Most of cell surface proteins are post-translationally modified by the attachment of different sugar- and/or lipid-residues (glycosylation, isoprenylation) for transport and/or stabilisation of the correct conformation of proteins within the cell membrane.

The stability and processing of CLCA2 was analysed in the presence of the glycosylation inhibitor Tunicamycin and the isoprenylation inhibitor Perillic acid, utilising stable CLCA2-expressing T47D Tet-on cell lines (inducible expression system). Cells were treated with increased concentrations of the corresponding inhibitors and protein lysates analysed on Western blot, by detection of CLCA2 protein via the HA-tag, using an anti-HA-antibody. As seen in Figure 52 the stability of CLCA2 seems to be dependent on its post-translational glycosylation. Furthermore, glycosylation influences proteolytic processing: At a concentration of 0.5 µg/µl of Tunicamycin no proteolytic processing is observed anymore.

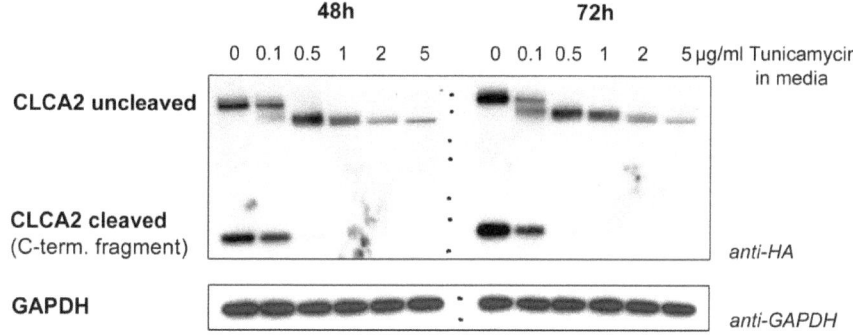

Figure 52. CLCA2 stability and post-translational proteolytic processing of CLCA2 in the presence of the glycosylation inhibitor Tunicamycin
300,000 cells of CLCA2-expressing T47D Tet-on clones #3 were seeded into 6-well-plates and treated with 0, 0.1, 0.5, 1, 2, 5 µg/ml Tunicamycin in media. CLCA2 expression was induced with Doxycycline. Protein lysates were harvested 48 h post induction. For Western blot 50 µg protein was loaded per lane; an anti-HA-tag antibody was used for detection of HA-tagged CLCA2. GAPDH served as loading control.

Analyses with several glycosylation site prediction software tools resulted in prediction of the putative glycosylation sites N150 and N822. However, mutation of these amino acids by exchanging Asparagine with Alanine did not influence the proteolytic processing of CLCA2 (Figure 53). Therefore, N150 and N822 seem not to be the major glycosylation sites in CLCA2.

However, N822 is located within the C-terminal fragment of CLCA2 and interestingly, a point mutation at this position influences the proteolytic processing resulting in a shorter C-terminal fragment on Western blots (Figure 53).

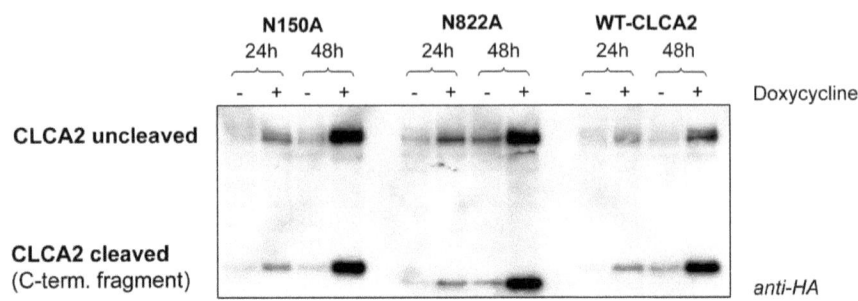

Figure 53. Mutations of putative glycosylation sites N150A and N822A of CLCA2 and their effects on its stability and proteolytic processing
150,000 T47D Tet-on cells were seeded into 6-well-plates and transiently transfected with constructs of pTRE2pur expression vector, expressing either HA-tagged wt-CLCA2 or bearing one of the following mutations: N150A, N822A. Expression was non-induced (-) and induced (+) with Doxycycline 24 h post transfection. Protein lysates were harvested 24 h and 48 h post induction. For Western blot 50 µg protein was loaded per lane; an anti-HA-tag antibody was used for detection of HA-tagged wt/mutant CLCA2.

In contrast to inhibition of glycosylation (Figure 52), isoprenylation does not play an important role for stability or efficient proteolytic cleavage of CLCA2 (Figure 54).

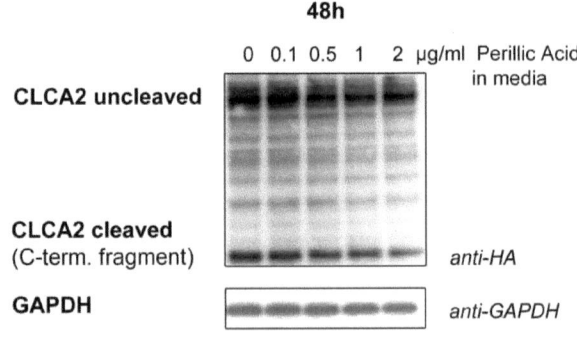

Figure 54. CLCA2 stability and post-translational proteolytic processing of CLCA2 in the presence of the isoprenylation inhibitor Perillic acid
300,000 cells of CLCA2-expressing T47D Tet-on clones #3 were seeded into 6-well-plates and treated with 0, 0.1, 0.5, 1, 2, 5 µg/ml Perillic acid in media. CLCA2 expression was induced with Doxycycline. Protein lysates were harvested 48 h post induction. For Western blot 50 µg protein was loaded per lane; an anti-HA-tag antibody was used for detection of HA-tagged CLCA2. GAPDH served as loading control.

Integrin β4

Abdel-Ghany found Integrin β4 to be coprecipitated with CLCA2. He also described a strong correlation between the level of Integrin β4 and adhesion of HEK293 cells expressing CLCA2 and a correlation *in vivo* between Integrin β4 expression in tumourigenic cell lines and the number of lung metastases detected in nude mice [70]. Upregulated Integrin β4 levels were also found in microarray studies of lung cancer metastases [74].

It could be shown that induction of CLCA2 expression in CLCA2-expressing T47D Tet-on clones #3 leads to enhanced Integrin β4 expression (Figure 55).

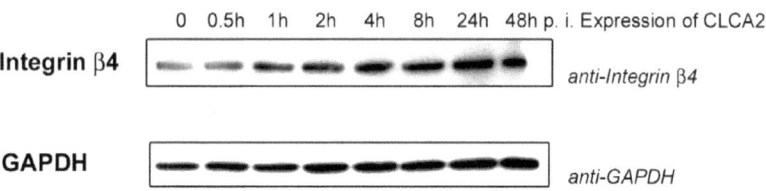

Figure 55. Expression of Integrin β4 upon induction of CLCA2 expression
Expression of CLCA2 was induced in CLCA2-expressing T47D Tet-on clone #3 by addition of Doxycycline.
Protein lysates were harvested at indicated time points post induction. CLCA2 expression of lysates was confirmed before use. For Western blot 50 µg protein was loaded per lane; an anti-Integrin β4 antibody was used for detection of Integrin β4. GAPDH served as loading control.

Due to the lack of phospho-specific antibodies towards Intgerin β4, no conclusion could be drawn whether Integrin β4 becomes not only upregulated, but also activated upon induction of CLCA2.

Integrin-Mediated Cell Adhesion Assay

Based on the data from Abdel-Ghany et al. [70] and to gain a more detailed understanding about the specific induction of integrins upon induced CLCA2 expression, the "Integrin-Mediated Cell Adhesion Arrays" (Chemicon) were used. (For details see section "Material and Methods".)

Especially the protein family of integrins (cell surface receptors) is involved in cell attachment to proteins of the extracellular matrix. Integrins therefore play an important role during metastasis, regulating not only dissemination but also homing of metastasising tumour cells. The "Integrin-Mediated Cell Adhesion Arrays" from Chemicon are efficient tools to screen cell surface protein profiles. These kits are composed of stripwells with each well pre-coated with an antibody detecting an individual integrin or extracellular matrix protein and one negative control well coated with bovine serum albumin (BSA). Utilising the established inducible expression system for CLCA2, it was possible to screen for induction of α- and β-integrins in dependence on induction of wt or mutant CLCA2 expression in T47D Tet-on cells. As a control T47D cells with "empty" expression were used. As seen in Figure 56 out of the α-integrin family no specific induction could be detected, whereas for β-integrin family member B2 a specific induction depending on wt-CLCA2 induced expression could be demonstrated for Clone #4 (Figure 57).

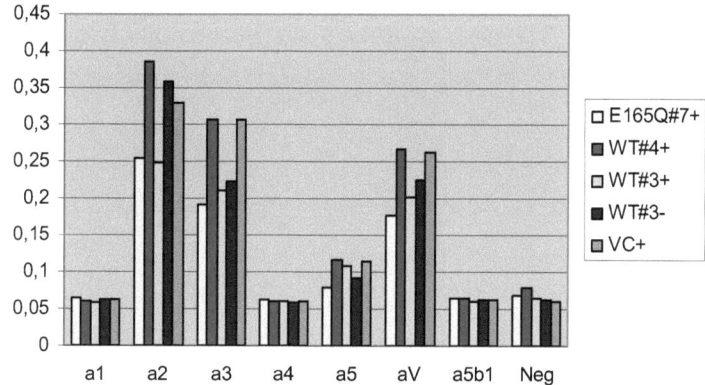

Figure 56. α-integrin-mediated cell adhesion of CLCA2-expressing and non-expressing cells
Expression of wt-CLCA2 was induced in T47D Tet-on clone #4, T47D Tet-on clone #3; expression of mutant-CLCA2 (E165Q) was induced in T47D Tet-on clone #7. As control served non-induced T47D Tet-on clone #3 and T47D cells with the empty vector.
Lysates were tested in the "Alpha-Integrin-Mediated Cell Adhesion Array" from Chemicon.

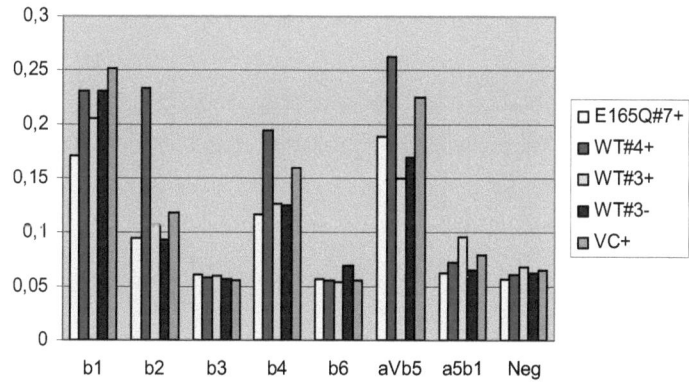

Figure 57. β-integrin-mediated cell adhesion of CLCA2-expressing and non-expressing cells
Expression of wt-CLCA2 was induced in T47D Tet-on clone #4, T47D Tet-on clone #3; expression of mutant-CLCA2 (E165Q) was induced in T47D Tet-on clone #7. As control served non-induced T47D Tet-on clone #3 and T47D cells with the empty vector.
Lysates were tested in the "Beta-Integrin-Mediated Cell Adhesion Array" from Chemicon.

3D-In Vitro Carcinoma Assay: Spheroids

For a better simulation of the situation in a tumour, CLCA2-expressing clones were not only grown in 2D cell culture, but also in multicellular tumour spheroids, which are *in vitro* 3D models that simulate malignant cell contacts within a tumour. Multicellular tumour spheroids can for instance also be used to evaluate tumour response to therapeutic agents. In this approach the *in vitro* 3D model was used for analysis of the influence of CLCA2 expression on cell growth and viability in a 3D model.

For this purpose 250 cells, each of the various T47D-CLCA2 clones were cultivated with methylcellulose and transferred into 96-well-plates for suspension culture. The methylcellulose causes cells to grow in three-dimensional cultures as opposed to mono-layer, forming a spheroid.

After forming stable spheroids, they were transferred to a collagen gel or Matrigel (Becton Dickinson) and grown until the spheroids became necrotic due to increased volume and the lack of supply with nutrients and oxygen in the centre of the 3D spheroid, as angiogenesis-induced vessels in a tumour would do.

After 3, 5, 7, and 10 days in the gel, spheroids were monitored under the microscope. Growth was estimated by measurement of the area of spheroids with AxioVision Software (Zeiss). To analyse the apoptotic status of cells within a spheroid, cells were stained with the intercalating and fluorescent agent Propidium iodide that pervades cell membranes of dead cells only.

Before forming spheroids, stable CLCA2 expression of T47D Tet-on clones was confirmed on Western blot (data not shown).

The following cell clones were used in this study:
- wt-CLCA2-expressing T47D Tet-on clone #3 (in induced/non-induced status)
- wt-CLCA2-constitutively-expressing T47D clone #47
- mutant-CLCA2(E165Q)-expressing T47D Tet-on clone #7 (in induced status)
- control cells, which stably carry the "empty" expression vector pTRE2pur (no CLCA2 expression) and which were pseudo-induced with Doxycycline to search for any specific Doxycycline-dependent effect.

Spheroids of non-induced wt-CLCA2-expressing T47D Tet-on clone #3 are shown in Figure 58 and can be directly compared with spheroids of the same clone, but in an induced status, as shown in Figure 59. Exemplarily, two representative pictures are shown for the time points 3, 5, 7, and 10 days growth of spheroids in Matrigel. No distinct difference could be observed between induced and non-induced clone #3, but in contrast a pronounced difference in area, representative for cell growth and apoptotic status of cells within the spheroid was found, as compared to the constitutive expressing clone #47 (Figure 60). It seems that constitutive expression triggers cell growth in a way that these spheroids become considerably bigger from day 3 on in Matrigel. It is also of interest that in spheroids of clone #47 considerably more apoptotic cells were found than in clone #3, where CLCA2 expression was induced only when spheroids were transferred to the Matrigel. Therefore, constitutive CLCA2 expression during spheroid-formation lead to generation of larger spheroids and subsequently to more apoptosis activity due to enhanced cell growth, supporting previous findings that CLCA2 essentially contributes to proliferation and/or viability of tumour cells.

Control cells, which stably carry the "empty" expression vector pTRE2pur and which were "pseudo-induced" with Doxycycline, are shown in Figure 61 and look very similar to non-induced clone #3 (Figure 58). Cells expressing mutant CLCA2 protein (E165Q), which are exhibiting a deficient proteolytic processing, were undistinguishable from other spheroids (Figure 62). Apparently proteolytic processing did not have any influence on spheroid formation or other properties of spheroids concerning cell growth or apoptotic status of cells within a spheroid. The same holds true for other CLCA2 expressed mutants (data not shown).

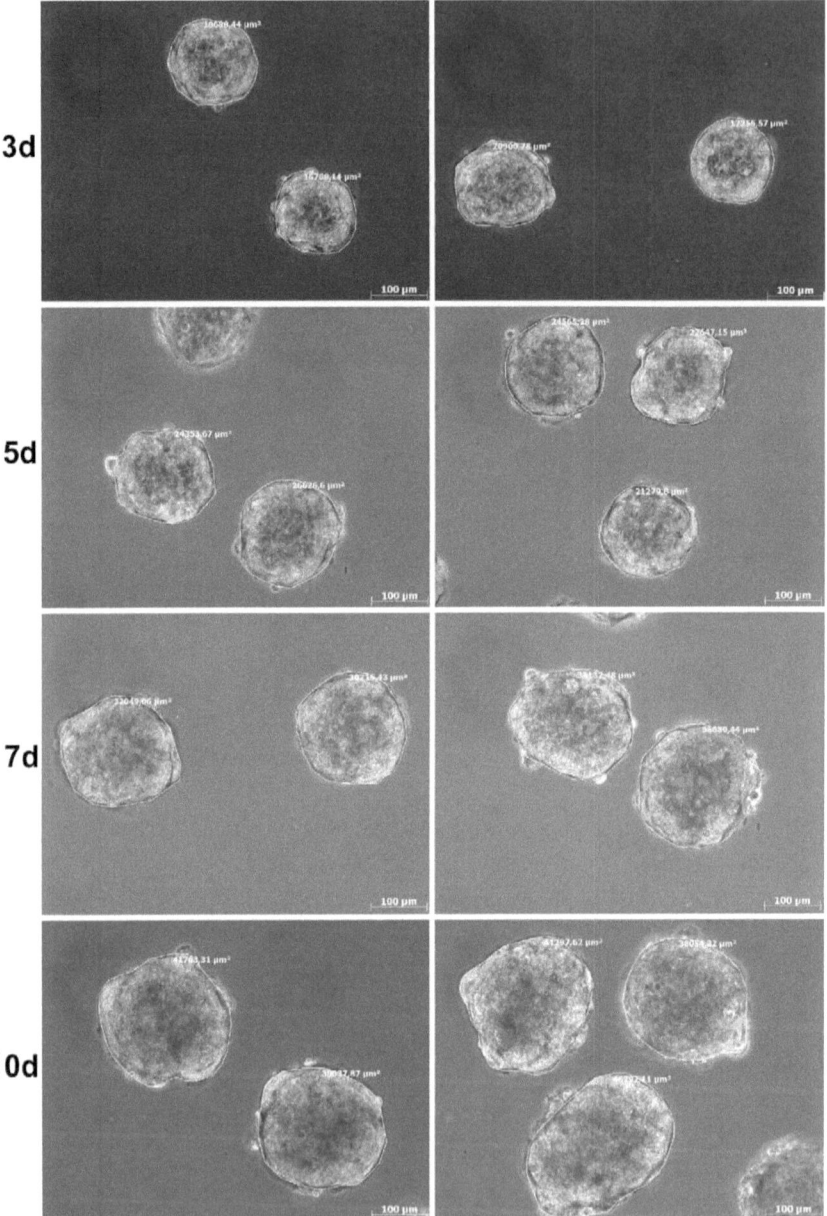

Figure 58. Spheroids of wt-CLCA2-expressing T47D Tet-on clone #3 – non-induced
Exemplarily two pictures are shown after growth for 3 d, 5 d, 7 d, and 10 d in Matrigel.
Red: Propidium iodide-stained apoptotic cells.

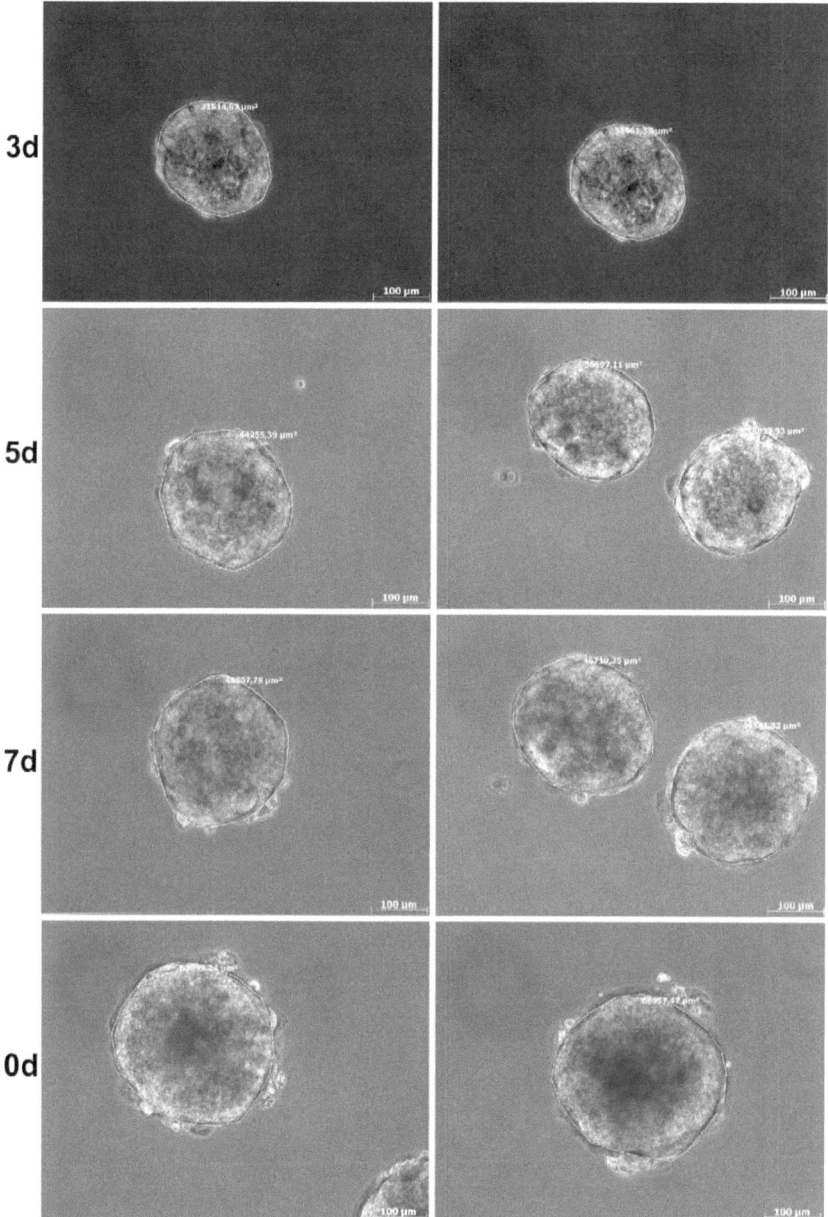

Figure 59. Spheroids of wt-CLCA2-expressing T47D Tet-on clone #3 – induced
Exemplarily two pictures are shown after growth for 3 d, 5 d, 7 d, and 10 d in Matrigel.
Red: Propidium iodide-stained apoptotic cells.

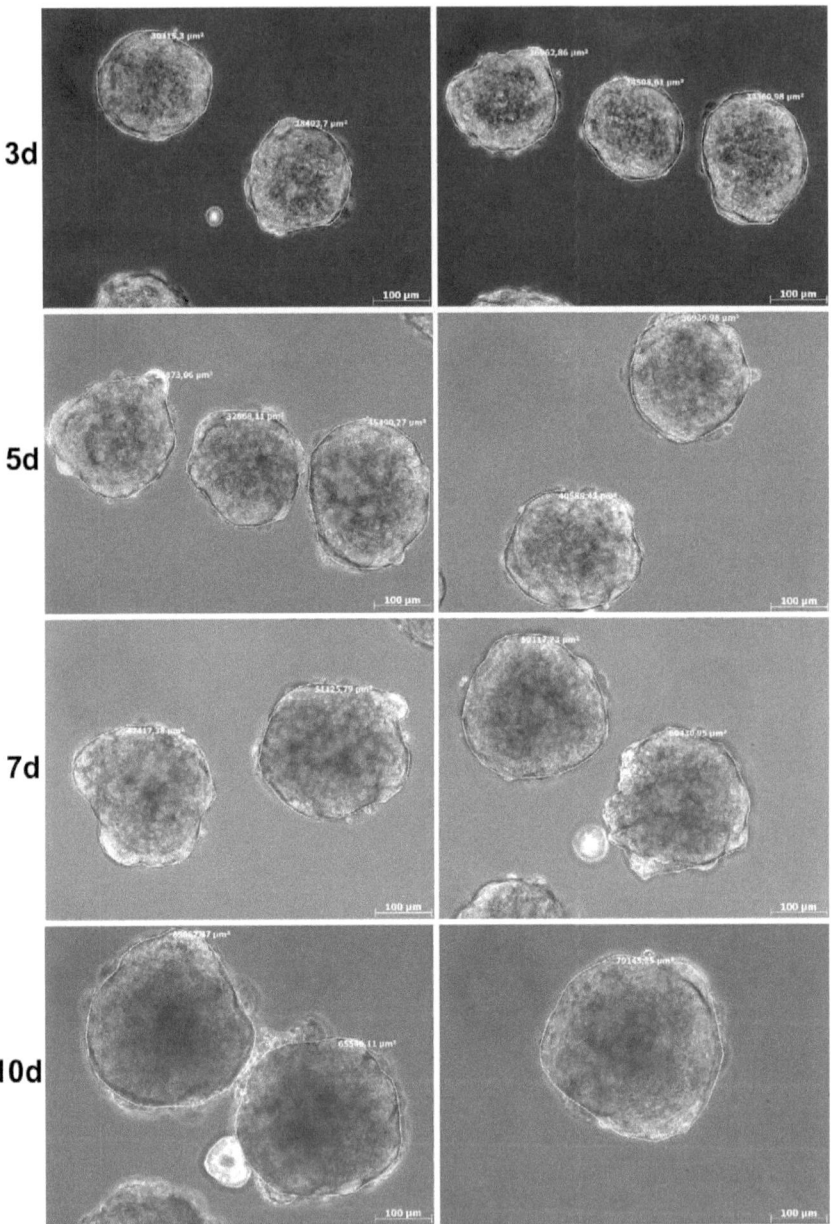

Figure 60. Spheroids of wt-CLCA2-constitutively-expressing T47D clone #47
Exemplarily two pictures are shown after growth for 3 d, 5 d, 7 d, and 10 d in Matrigel.
Red: Propidium iodide-stained apoptotic cells.

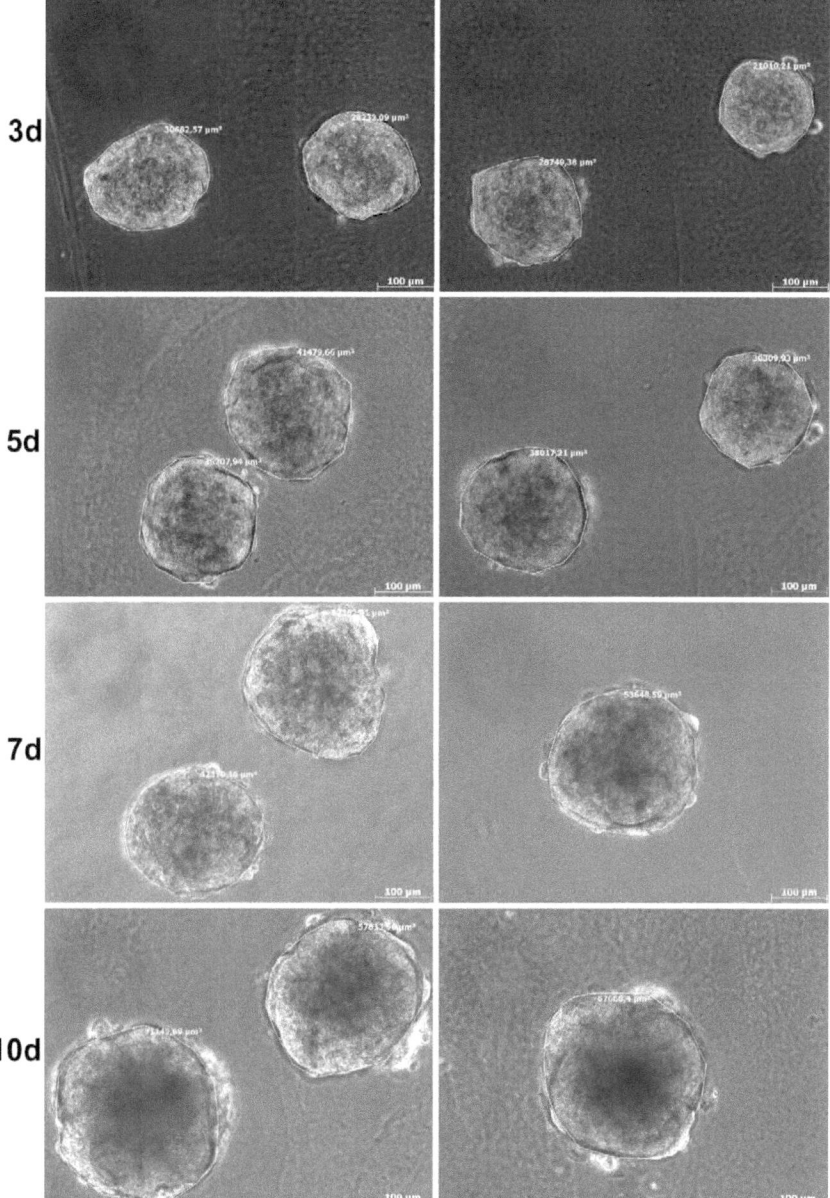

Figure 61. Spheroids of control cells (T47D Tet-on cells with empty vector, Doxycycline-treated)
Control cells stably carrying the "empty" expression vector pTRE2pur were pseudo-induced with Doxycycline for analysis of any Doxycycline effect. Exemplarily two pictures are shown after growth for 3 d, 5 d, 7 d, and 10 d in Matrigel. Red: Propidium iodide-stained apoptotic cells.

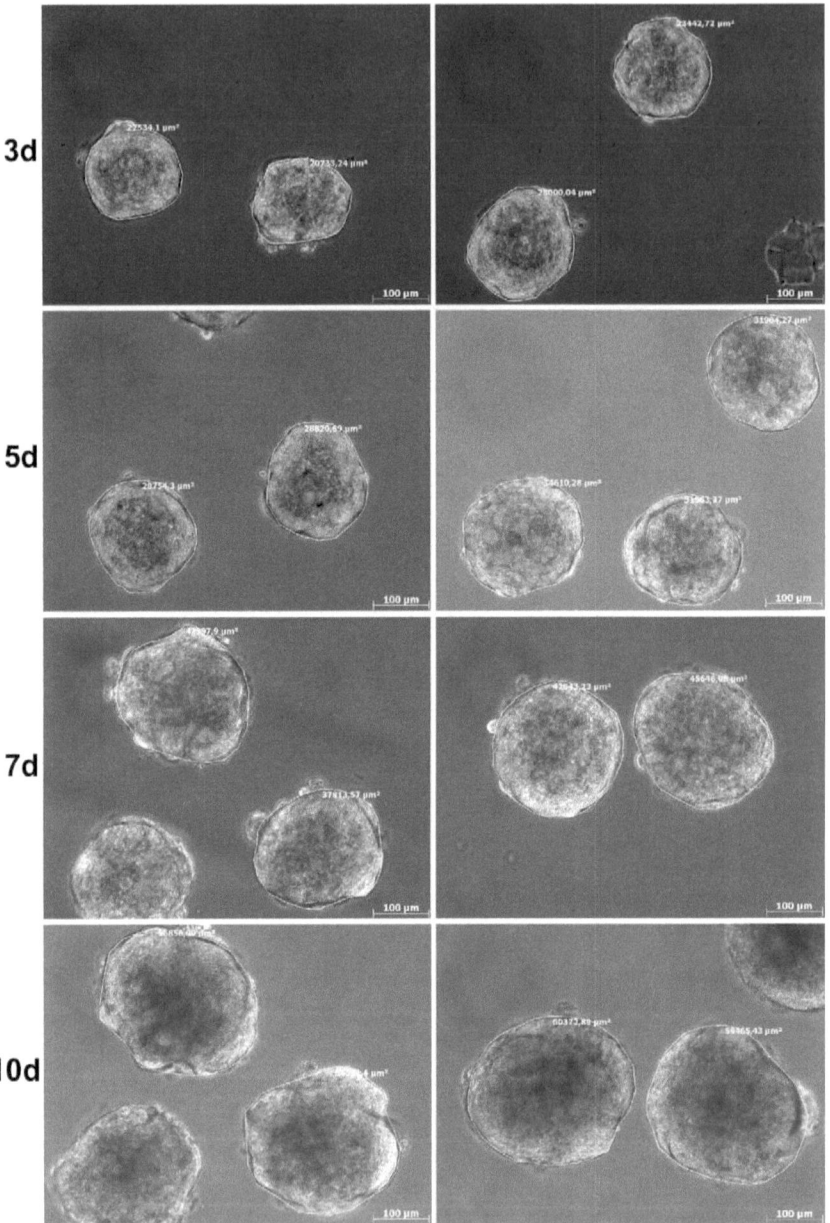

Figure 62. Spheroids of mutant-CLCA2(E165Q)-expressing T47D Tet-on clone #7 – induced
Exemplarily two pictures are shown after growth for 3 d, 5 d, 7 d, and 10 d in Matrigel.
Red: Propidium iodide-stained apoptotic cells.

Growth Curve of CLCA2-Expressing Clones (T47D Tet-on): 2D versus 3D

As it could be shown, that cell growth in the spheroid 3D-model is enhanced in CLCA2-consitutively-expressing T47D clone #47 compared to induced or non-induced CLCA2-expressing T47D Tet-on clones #3. Cell growth of these clones was also compared in 2D-tissue culture. Therefore, 2D-growth curves were generated, based on cell numbers of cells which had been trypsinised and counted after 3-10 days in 2D-cell culture. Interestingly, in 2D no distinct difference between non-induced or induced clone #3 could be observed.

Next, a 3D-growth curve was generated by measurement of the area of various spheroids, which was calculated with AxioVision Software after 3-10 days in Matrigel. In a statistically meaningful setting the relative cell growth of 2D- (counting viable cells) with that of 3D-cultures (determination of area [μm^2]; Figure 63) were compared over a period of 3 to 10 days. Although, the relative cell growth of induced versus non-induced clone #3 did not show a difference in 2D-cultures, it indeed exhibited a pronounced difference (> 2.5-fold) between 2D- and 3D-culturing. In contrast to inducible clone #3, the constitutive clone #47 showed already an enhanced relative cell growth in 2D-culture, which was similar to that found in 3D-cultures. Spheroids of inducible cell clones expressing other CLCA2-mutants were also analysed in this assay, but none of them showed a statistically relevant different behaviour in growth, when compared to inducible CLCA2 clones (data not shown).

In general, the 3D-model is more sensitive than 2D-cultures to show differences in cell proliferation, and it allows to demonstrate the impact of CLCA2 on the three-dimensional growth of tumour cells.

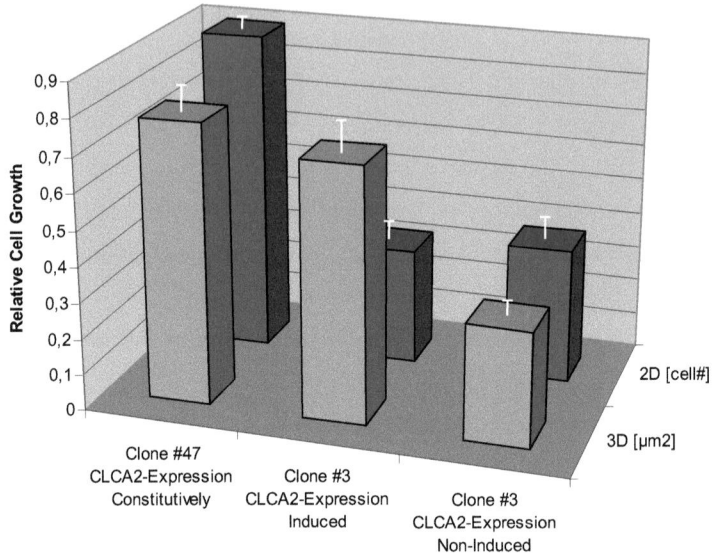

Figure 63. Relative cell growth in 2D and 3D of T47D clones with constitutive, induced, and non-induced CLCA2 expression
Left: CLCA2-constitutively-expressing T47D clone
Middle: CLCA2-expressing T47D Tet-on clones #3 – induced
Right: CLCA2-expressing T47D Tet-on clones #3 – non-induced
For 2D-growth curve, cells were trypsinised and counted after 3-10 days in 2D-cell culture.
For 3D-growth curve, area of spheroids was estimated with AxioVision Software after 3-10 d in Matrigel.
The value of area is the average of all measured spheroids (5 up to 11 spheroids per time point).

Affymetrix Microarray Studies

To study the effects of overexpression of CLCA2 on gene expression, expression profiling using the Affymetrix GeneChip Human Genome U133 Array was performed by testing stable cell clones with inducible expression of CLCA2.

First, different wt-CLCA2-expressing T47D Tet-on clones (Figure 41) were tested for CLCA2 expression both, on the protein level performing Western blots and on the transcript level performing real-time quantitative PCR (RT-qPCR). Freshly thawn clones #3, #4 and #46 still proved as the best inducible cell lines as shown previously (Figure 41). These clones exhibit the most efficient induction profile (Figure 64).

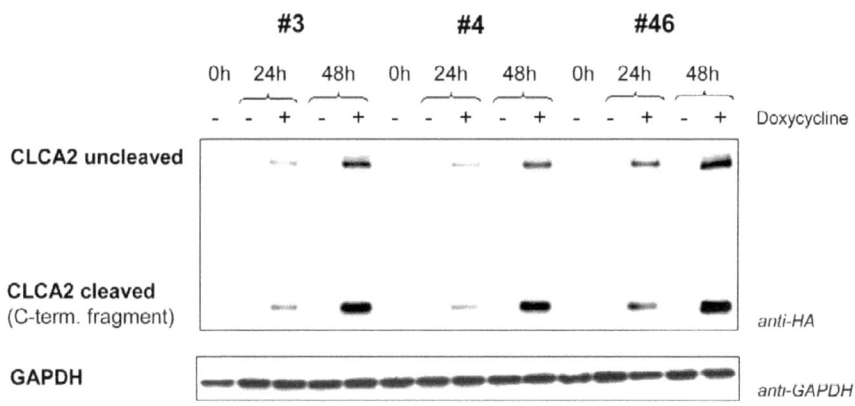

Figure 64. CLCA2 protein expression of inducible CLCA2-expressing T47D Tet-on clones used for Affymetrix microarray studies
300,000 cells of CLCA2-expressing T47D Tet-on clone #3, #4, and #46 were seeded each into 6-well-plates and induced with Doxycycline. Protein lysates were harvested 24 h and 48 h post induction. For Western blot 50 µg protein was loaded per lane; an anti-HA-tag antibody was used for detection of HA-tagged CLCA2. GAPDH served as loading control.

Expression of CLCA2 transcript in clone #3 was monitored by performing RT-qPCR with cDNA samples taken at different time points. As shown in Figure 65 the CLCA2 transcript could only be detected in induced lysates. CLCA2 transcript level increased between 8 h and 24 h post induction. In non-induced lysates the CLCA2 transcript stayed within the background level of control cells (Figure 65).

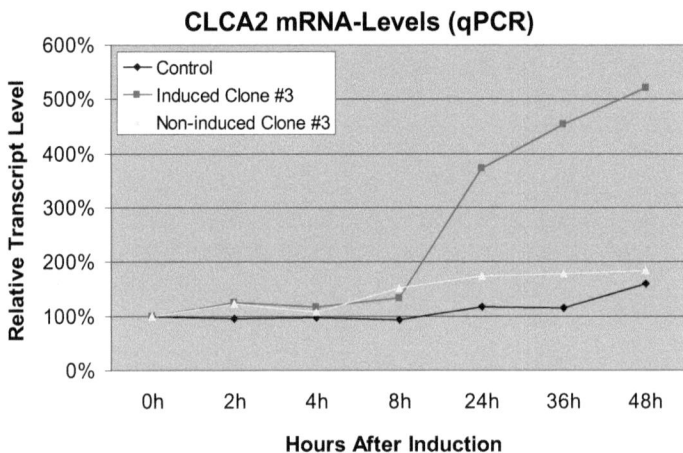

Figure 65. CLCA2 mRNA levels of induced and non-induced CLCA2-expressing T47D Tet-on clone #3 versus control cells (T47D Tet-on cells with empty vector, Doxycycline-treated)
300,000 cells were seeded into 6-well-plates and induced with Doxycycline. RNA lysates were harvested 24 h post induction and transcribed into cDNA. For detection of the CLCA2 transcript by RT-qPCR a CLCA2-specific primer pair (see "Material and Methods") was used in a duplex assay together with the internal control β-2 microglobulin for normalisation.

After testing primer pairs for various housekeeping genes in these lysates (data not shown), β-2-microglobulin was chosen for use as the most robust internal reference for normalisation of CLCA2 transcript levels.

Protein lysates were prepared at various time points and stored at -80°C. cDNA preparations of clones #4 and #46 were analysed on Affymetrix GeneChip Human Genome U133 Array (for details see "Material and Methods"). Expression profiles are shown in Figure 66.

Blue bars indicate CLCA2-trancript levels in non-induced cells and red bars CLCA2 transcript levels in induced cells, respectively. In all three clones induction of CLCA2 expression can be detected at 24 h and 48 h after treatment with Doxycycline. In contrast to other clones, clone #46 exhibits already elevated CLCA2 levels without induction. However, the relative induction range is similar compared to clone #3 and #4. Clone #46 obviously contains a "leaky" expression control for CLCA2.

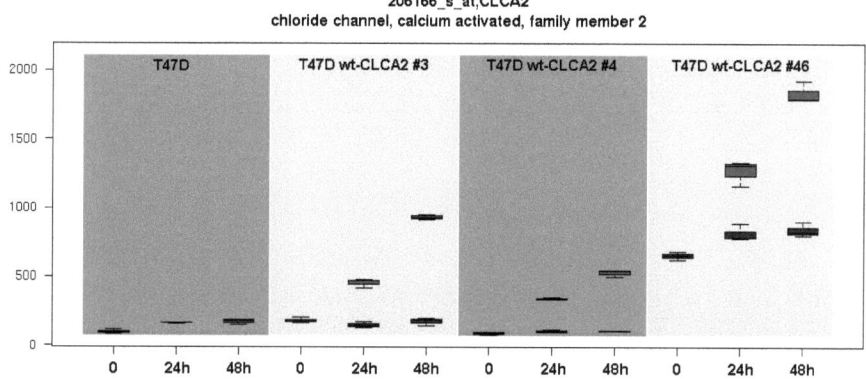

Figure 66. CLCA2 mRNA levels of inducible CLCA2-expressing T47D Tet-on clones #3, #4 and #46 compared with T47D (control) on Affymetrix GeneChip Human Genome U133 Array
The MAS5-normalised expression levels of CLCA2 (206166_s_at) mRNA is shown in box-plots. Light grey box-plots indicate CLCA2 transcript levels of cells with induced CLCA2 expression, dark grey box-plots represent CLCA2 transcript levels of non-induced cells. The vertical centre line in the box indicates the median; the box itself represents the interquartile range (IQR) between the first and third quartiles.

RESULTS

In a next step, genes whose induction is dependent on the presence/induction of CLCA2 should be identified; i.e. these are transcripts which are induced in CLCA2-inducibly-expressing cell lines, but not expressed in non-induced or control cells ("empty" vector bearing cells). Selected top hits are shown in Table 2. To reduce complexity of data this "correlation analysis" was only performed for clone #3.

Probeset	Gene Symbol	Gene Title	Chromosomal Location	Correlation Factor (r)
206166_s_at	CLCA2	chloride channel, calcium activated, family member 2	1p31-p22	1
212134_at	PHLDB1	pleckstrin homology-like domain, family B, member 1	11q23.3	0,98308
223212_at	ZDHHC16	zinc finger, DHHC-type containing 16	10q24.1	0,98276
229399_at	C10orf118	chromosome 10 open reading frame 118	10q25.3	0,98070
202733_at	P4HA2	procollagen-proline, 2-oxoglutarate 4-dioxygenase (proline 4-hydroxylase), alpha polypeptide II	5q31	0,97973
200974_at	ACTA2	actin, alpha 2, smooth muscle, aorta	10q23.3	0,97693
215120_s_at	SAMD4	sterile alpha motif domain containing 4	14q22.2	0,97123
209922_at	BRAP	BRCA1 associated protein	12q24	0,96837
1557104_at	NA	Zinc finger protein	NA	0,96594
203408_s_at	SATB1	special AT-rich sequence binding protein 1 (binds to nuclear matrix/scaffold-associating DNA's)	3p23	0,95899
1552621_at	POLR2J2	DNA directed RNA polymerase II polypeptide J-related gene	7q11.22	0,95825
203050_at	TP53BP1	tumor protein p53 binding protein, 1	15q15-q21	0,95779
233750_s_at	C1orf25	chromosome 1 open reading frame 25	1q25.2	0,95680
232754_at	NA	NA	NA	0,95256
223629_at	PCDHB5	protocadherin beta 5	5q31	0,95240
1556339_a_at	NA	NA	NA	0,95147
224321_at	TMEFF2	transmembrane protein with EGF-like and two follistatin-like domains 2 /// transmembrane protein with EGF-like and two follistatin-like domains 2	2q32.3	0,95135
201295_s_at	WSB1	WD repeat and SOCS box-containing 1	17q11.1	0,95100
208589_at	TRPC7	transient receptor potential cation channel, subfamily C, member 7	5q31.1	0,94950
1554895_a_at	RHBDL2	rhomboid, veinlet-like 2 (Drosophila)	1p34.3	0,94919
211762_s_at	KPNA2	karyopherin alpha 2 (RAG cohort 1, importin alpha 1) /// karyopherin alpha 2 (RAG cohort 1, importin alpha 1)	17q23.1-q23.3	0,94716
235756_at	NA	CDNA FLJ26187 fis, clone ADG04782	NA	0,94685
224280_s_at	FAM54B	family with sequence similarity 54, member B	1p36.11	0,94434
207904_s_at	LNPEP	leucyl/cystinyl aminopeptidase	5q15	0,94389
1553740_a_at	IRAK2	interleukin-1 receptor-associated kinase 2	3p25.3	0,94332

Table 2. "Correlation Analysis" on expression pattern of microarray studies
The Table lists an extract of genes with a high correlation factor showing an induction upon induced expression of CLCA2 in clone #3 after 0, 24 h and 48 h. A correlation factor of 1,000 represents the induction of CLCA2 (probeset 206166_s_at). The right column refers to the respective correlation factor r.

As clone #4 and #46 had also been analysed on Affymetrix GeneChips in parallel, the selected top-correlating probesets (genes) listed in Table 2 could be manually compared with the corresponding expression pattern of clone #4 and #46. Those genes showing a similar expression profile in clone #4 and #46 were selected and subjected to RT-qPCR to confirm the expression pattern by an independent method. Individual results are shown in Figure 67 to Figure 83, always comparing the respective expression profile from the Affymetrix GeneChip analysis with the corresponding one from RT-qPCR analysis.

SFI1

SFI1 homolog, spindle assembly associated (yeast) is involved in initiation of centrosome duplication. SFI1 has been identified as a binding partner for the calcium-binding protein centrin suggesting a role in the dynamic behavior of centrosomes. SFI1 binds to multiple centrin molecules and the complex forms calcium sensitive contractile fibers that function to reorient centrioles and alter centrosome structure [96]. So far no correlation to tumour cell growth has been reported. However, according to our *in silico* analysis SFI1 is found upregulated in many tumors, with highest expression levels found in various B-cell lymphomas and B-cell leukemias (data not shown).

Results from Affymetrix microarray analysis indicating an upregulation of SFI1 upon induction of CLCA2 expression could be confirmed in RT-qPCR as shown in Figure 67.

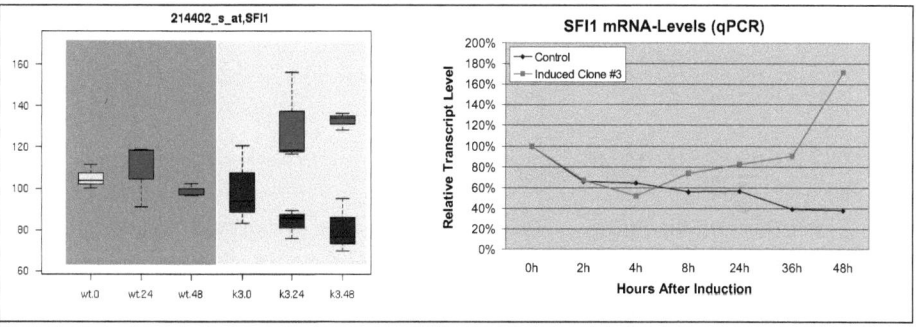

Figure 67. SFI1 mRNA levels in induced CLCA2-expressing T47D Tet-on clone #3 versus control
Left: Affymetrix microarray studies using Affymetrix GeneChip Human Genome U133 Array
The MAS5-normalised expression levels of SFI1 mRNA are shown in box-plots. The vertical centre line in the box indicates the median; the box itself represents the interquartile range (IQR) between the first and third quartiles. Whiskers extend to 1.5 times the IQR. In the left dark grey field, expression levels in control cells (T47D Tet-on) are shown. In the right light grey field, expression levels in T47D Tet-on clone #3 (k3) are represented.
Light grey box-plots indicate SFI1 transcript levels of cells with induced CLCA2 expression; dark grey box-plots represent SFI1 transcript levels of non-induced cells. The white box-plot represents the transcript level in control cells at the time point of induction. 0, 24 and 48 indicate the time points (h) of lysate preparation post induction
Right: RT-qPCR in duplex assay with β-2-microglobulin as internal reference for normalisation
Normalised SFI1 mRNA levels of induced clone #3 are shown in relation to mRNA levels in control cells (T47D Tet-on with empty expression vector).

TP53BP1

Alteration of the tumour protein TP53BP1 (p53 binding protein 1), during skin carcinogenesis is associated with genomic instability [97]. The expression pattern of TP53BP1 from microarray analysis is shown in Figure 68, left side suggesting an overexpression of TP53BP1 if CLCA2 expression is induced. However, higher levels of TP53BP1 could also be detected in control cells after 48 h of treatment with Doxycycline. In addition, overexpression of TP53BP1 could also not be confirmed with RT-qPCR (Figure 68, right side). Therefore, expression of TP53BP1 has to be regarded as a non-specific induction. This example supports the application of an independent method for validation of expression profiles.

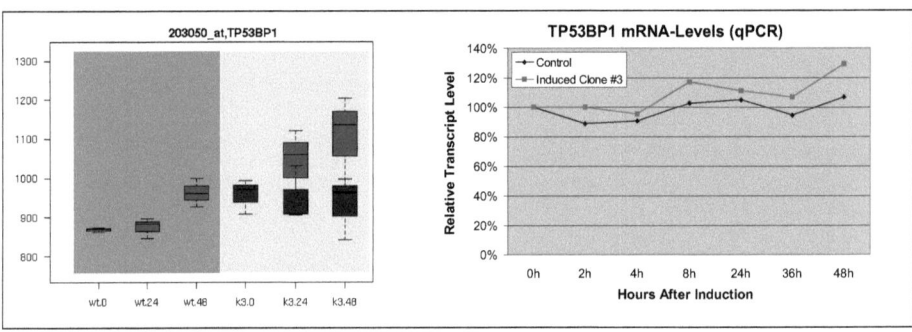

Figure 68. TP53BP1 mRNA levels in induced CLCA2-expressing T47D Tet-on clone #3 versus control
Left: Affymetrix microarray studies using Affymetrix GeneChip Human Genome U133 Array
The MAS5-normalised expression levels of TP53BP1 mRNA are shown in box-plots. The vertical centre line in the box indicates the median; the box itself represents the interquartile range (IQR) between the first and third quartiles. Whiskers extend to 1.5 times the IQR. In the left dark grey field, expression levels in control cells (T47D Tet-on) are shown. In the right light grey field, expression levels in T47D Tet-on clone #3 (k3) are represented.
Light grey box-plots indicate TP53BP1 transcript levels of cells with induced CLCA2 expression; dark grey box-plots represent TP53BP1 transcript levels of non-induced cells. The white box-plot represents the transcript level in control cells at the time point of induction. 0, 24 and 48 indicate the time points (h) of lysate preparation post induction.
Right: RT-qPCR in duplex assay with β-2-microglobulin as internal reference for normalisation
Normalised TP53BP1 mRNA levels of induced clone #3 is shown in relation to mRNA levels in control cells (T47D Tet-on with empty expression vector).

CDIPT

CDP-diacylglycerol-inositol 3-phosphatidyltransferase (phosphatidyinositol synthase) is a gene whose inhibition causes small cell lung carcinoma cells to arrest in G1 [98]. It further has been suggested to be involved in oral carcinogenesis [99]. Inostamycin, an inhibitor of CDIPT, led to a decrease in cyclin D1 mRNA and protein expression accompanied by suppression of phosphorylated retinoblastoma susceptibility gene product (pRB-P) levels. Inostamycin might be a useful agent for tumour dormant cytostatic therapy for oral SCC.

Overexpression of CDIPT was found in microarray studies and was confirmed in RT-qPCR, as shown in Figure 69.

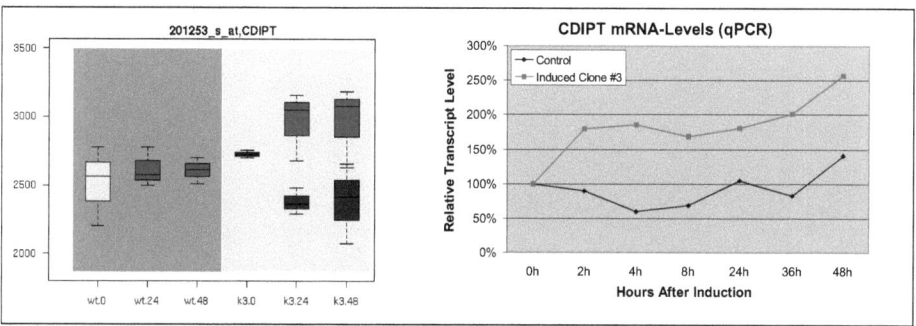

Figure 69. CDIPT mRNA levels in induced CLCA2-expressing T47D Tet-on clone #3 versus control

Left: Affymetrix microarray studies using Affymetrix GeneChip Human Genome U133 Array
The MAS5 normalised expression levels of CDIPT mRNA are shown in box-plots. The vertical centre line in the box indicates the median; the box itself represents the interquartile range (IQR) between the first and third quartiles. Whiskers extend to 1.5 times the IQR. In the left dark grey field, expression levels in control cells (T47D Tet-on) are shown. In the right light grey field, expression levels in T47D Tet-on clone #3 (k3) are represented.
Light grey box-plots indicate CDIPT transcript levels of cells with induced CLCA2 expression; dark grey box-plots represent CDIPT transcript levels of non-induced cells. The white box-plot represents the transcript level in control cells at the time point of induction. 0, 24 and 48 indicate the time points (h) of lysate preparation post induction.

Right: RT-qPCR in duplex assay with β-2-microglobulin as internal reference for normalisation
Normalised CDIPT mRNA levels of induced clone #3 are shown in relation to mRNA levels in control cells (T47D Tet-on with empty expression vector).

MLL2

Myeloid/lymphoid or mixed-lineage leukemia 2 gene plays a role in apoptosis and in alterations of cell adhesion. MLL2 is a member of the human MLL family, which belongs to a larger SET1 family of histone methyltransferases. MML2 is a transcriptional activator that induces the transcription of target genes by covalent histone modification. It appears to be involved in the regulation of adhesion-related cytoskeletal events, which might affect cell growth and survival [100]. Members of the MLL protein family have been found to cause acute leukemia. Deletion studies revealed that alterations of the CxxC domain (Zn-binding domain) disrupts its oncogenic potential [101]. Loss-of-function mutations within the coding sequence of NF-kappaB inhibitory molecules such as IkappaBalpha or p100 leads to constitutive NF-kappaB activation in haematological malignancies. Hut-78, a truncated form of p100, constitutively generates p52 and contributes to the development of T-cell lymphomas and was recently shown to induce MMP9 gene expression. Conversely, MMP9 gene expression is impaired in p52-depleted HUT-78 cells. Interestingly, MLL1 and MLL2 H3K4 methyltransferase complexes are tethered by p52 on the MMP9 but not on the IkappaBalpha promoter, and the H3K4 trimethyltransferase activity recruited on the MMP9 promoter is impaired in p52-depleted HUT-78 cells. Moreover, MLL1 and MLL2 are associated with Hut-78 in a native chromatin-enriched extract. This molecular mechanism recruits a H3K4 histone methyltransferase complex on the promoter of a NF-kappaB-dependent gene induces its expression and potentially the invasive potential of lymphoma cells harbouring constitutive activity of the alternative NF-kappaB-activating pathway [102].

Overexpression following CLCA2 induction could be observed in microarray studies and was confirmed by RT-qPCT (Figure 70).

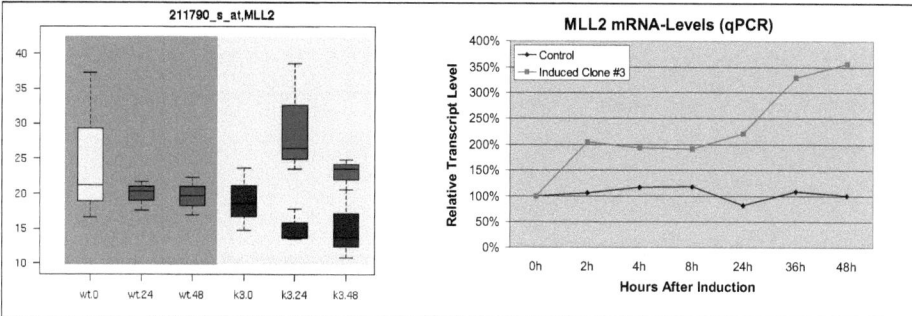

Figure 70. MLL2 mRNA levels in induced CLCA2-expressing T47D Tet-on clone #3 versus control
Left: Affymetrix microarray studies using Affymetrix GeneChip Human Genome U133 Array
The MAS5-normalised expression levels of MLL2 mRNA are shown in box-plots. The vertical centre line in the box indicates the median; the box itself represents the interquartile range (IQR) between the first and third quartiles. Whiskers extend to 1.5 times the IQR. In the left dark grey field, expression levels in control cells (T47D Tet-on) are shown. In the right light grey field, expression levels in T47D Tet-on clone #3 (k3) are represented.
Light grey box-plots indicate MLL2 transcript levels of cells with induced CLCA2 expression; dark grey box-plots represent MLL2 transcript levels of non-induced cells. The white box-plot represents the transcript level in control cells at the time point of induction. 0, 24 and 48 indicate the time points (h) of lysate preparation post induction.
Right: RT-qPCR in duplex assay with β-2-microglobulin as internal reference for normalisation
Normalised MLL2 mRNA levels of induced clone #3 are shown in relation to mRNA levels in control cells (T47D Tet-on with empty expression vector).

AFF4 (MCEF)

AF4/MR2 family member 4 (also known as MCEF), the newest member of the AF4 family of transcription factors involved in leukemia, is a positive transcription elongation factor-b-associated protein [103].
Its overexpression after CLCA2 induction shown on microarray arrays was confirmed by RT-qPCR (Figure 71).

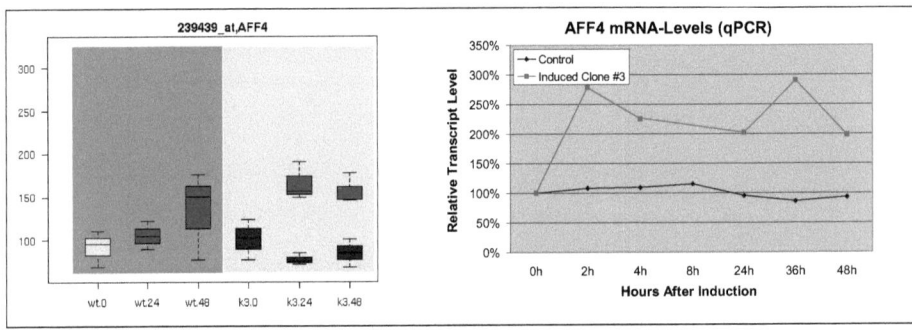

Figure 71. AFF4 mRNA levels in induced CLCA2-expressing T47D Tet-on clone #3 versus control
Left: Affymetrix microarray studies using Affymetrix GeneChip Human Genome U133 Array
The MAS5-normalised expression levels of AFF4 mRNA are shown in box-plots. The vertical centre line in the box indicates the median; the box itself represents the interquartile range (IQR) between the first and third quartiles. Whiskers extend to 1.5 times the IQR. In the left dark grey field, expression levels in control cells (T47D Tet-on) are shown. In the right light grey field, expression levels in T47D Tet-on clone #3 (k3) are represented.
Light grey box-plots indicate AFF4 transcript levels of cells with induced CLCA2 expression; dark grey box-plots represent AFF4 transcript levels of non-induced cells. The white box-plot represents the transcript level in control cells at the time point of induction. 0, 24 and 48 indicate the time points (h) of lysate preparation post induction.
Right: RT-qPCR in duplex assay with β-2-microglobulin as internal reference for normalisation
Normalised AFF4 mRNA levels of induced clone #3 are shown in relation to mRNA levels in control cells (T47D Tet-on with empty expression vector).

UNC84A

UNC84A belongs to the conserved family of SUN proteins. Human UNC84A (Sun1) is a homolog of C. elegans UNC-84, a protein involved in nuclear anchorage and migration. Targeting of UNC84A to the nuclear envelope revealed that the N-terminal 300 amino acids are crucial for efficient nuclear envelope localisation of UNC84A whereas the conserved C-terminal SUN domain is not required. Localisation and anchoring of UNC84A is not dependent on the lamin proteins, in contrast to what had been observed for C. elegans UNC-84. If UNC84A can also interact with cytoplasmic tracts of transmembrane proteins such as CLCA2 to contribute to ECM interaction/migration is not yet known [104].

The transcript is found upregulated both, by Affymetrix GeneChip analysis and RT-qPCR (Figure 72).

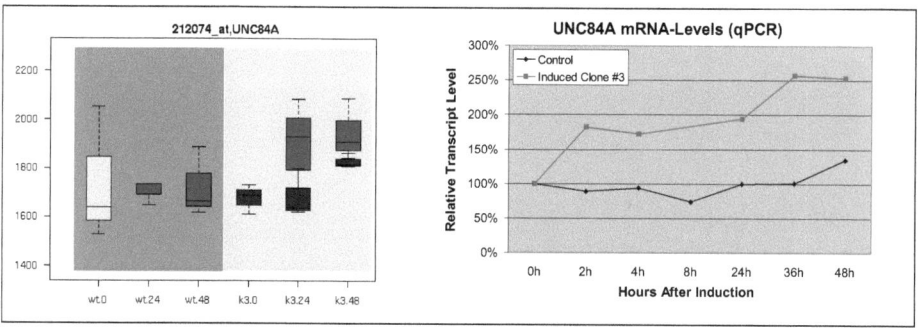

Figure 72. UNC84A mRNA levels in induced CLCA2-expressing T47D Tet-on clone #3 versus control
Left: Affymetrix microarray studies using Affymetrix GeneChip Human Genome U133 Array
The MAS5-normalised expression levels of UNC84A mRNA are shown in box-plots. The vertical centre line in the box indicates the median; the box itself represents the interquartile range (IQR) between the first and third quartiles. Whiskers extend to 1.5 times the IQR. In the left dark grey field, expression levels in control cells (T47D Tet-on) are shown. In the right light grey field, expression levels in T47D Tet-on clone #3 are represented.
Light grey box-plots indicate UNC84A transcript levels of cells with induced CLCA2 expression; dark grey box-plots represent UNC84A transcript levels of non-induced cells. The white box-plot represents the transcript level in control cells at the time point of induction. 0, 24 and 48 indicate the time points (h) of lysate preparation post induction.
Right: RT-qPCR in duplex assay with β-2-microglobulin as internal reference for normalisation
Normalised UNC84A mRNA levels of induced clone #3 are shown in relation to mRNA levels in control cells (T47D Tet-on with empty expression vector).

DTX2 (Deltex2)

The deltex null mutant indicates tissue-specific deltex-dependent Notch signaling in drosophila. It was further shown, that in drosophila deltex mediates suppressor of hairless-independent and late-endosomal activation of Notch signaling. In addition, the drosophila Nedd4-like protein suppressor of Deltex, Su(dx), has been characterised as a negative regulator of Notch receptor signaling, an intercellular signaling pathway of fundamental importance for multiple cell fate decisions [105;106].

Upregulation of DTX2 could be confirmed on the Affymetrix GeneChip and by RT-qPCR (Figure 73).

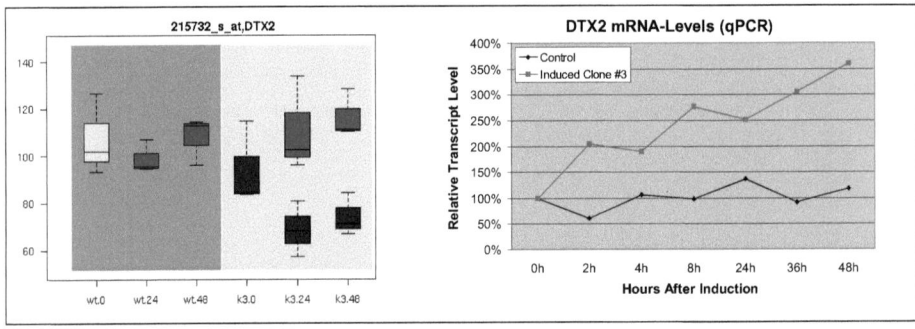

Figure 73. DTX2 mRNA levels in induced CLCA2-expressing T47D Tet-on clone #3 versus control
Left: Affymetrix microarray studies using Affymetrix GeneChip Human Genome U133 Array
The MAS5-normalised expression levels of DTX2 mRNA are shown in box-plots. The vertical centre line in the box indicates the median; the box itself represents the interquartile range (IQR) between the first and third quartiles. Whiskers extend to 1.5 times the IQR. In the left dark grey field, expression levels in control cells (T47D Tet-on) are shown. In the right light grey field, expression levels in T47D Tet-on clone #3 (k3) are represented.
Light grey box-plots indicate DTX2 transcript levels of cells with induced CLCA2 expression; dark grey box-plots represent DTX2 transcript levels of non-induced cells. The white box-plot represents the transcript level in control cells at the time point of induction. 0, 24 and 48 indicate the time points (h) of lysate preparation post induction.
Right: RT-qPCR in duplex assay with β-2-microglobulin as internal reference for normalisation
Normalised DTX2 mRNA levels of induced clone #3 are shown in relation to mRNA levels in control cells (T47D Tet-on with empty expression vector).

C10orf118

Chromosome 10 open reading frame 118 of unknown function was found upregulated after CLCA2-induction on the Affymetrix GeneChip and was confirmed by RT-qPCR (Figure 74).

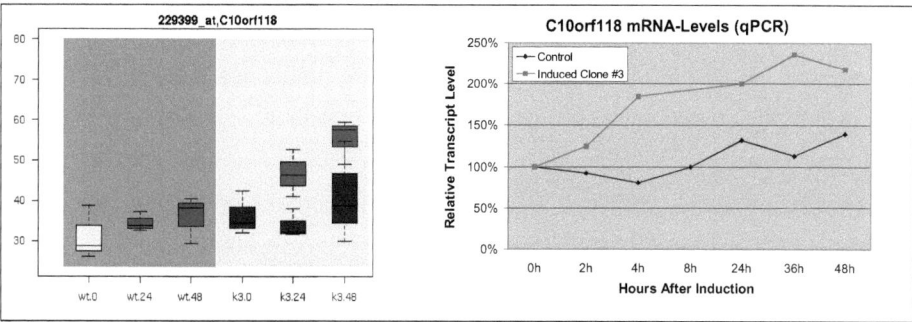

Figure 74. C10orf118 mRNA levels in induced CLCA2-expressing T47D Tet-on clone #3 versus control
Left: Affymetrix microarray studies using Affymetrix GeneChip Human Genome U133 Array
The MAS5-normalised expression levels of C10orf118 mRNA areshown in box-plots. The vertical centre line in the box indicates the median; the box itself represents the interquartile range (IQR) between the first and third quartiles. Whiskers extend to 1.5 times the IQR. In the left dark grey field, expression levels in control cells (T47D Tet-on) are shown. In the right light grey field, expression levels in T47D Tet-on clone #3 (k3) are represented.
Light grey box-plots indicate C10orf118 transcript levels of cells with induced CLCA2 expression; dark grey box-plots represent C10orf118 transcript levels of non-induced cells. The white box-plot represents the transcript level in control cells at the time point of induction. 0, 24 and 48 indicate the time points (h) of lysate preparation post induction.
Right: RT-qPCR in duplex assay with β-2-microglobulin as internal reference for normalisation
Normalised C10orf118 mRNA levels of induced clone #3 are shown in relation to mRNA levels in control cells (T47D Tet-on with empty expression vector).

SATB1

SATB1 is a genome organizer that tethers multiple genomic loci and recruits chromatin-remodelling enzymes to regulate chromatin structure and gene expression. It was shown that SATB1 is expressed by aggressive breast cancer cells and its expression level has high prognostic significance, independent of lymph-node status [107]. Knock-down of SATB1 in highly aggressive cancer cells altered the expression of lots of genes, reversing tumourigenesis by restoring breast-like acinar polarity and inhibiting tumour growth and metastasis *in vivo*. Conversely, ectopic SATB1 expression in non-aggressive cells led to gene expression patterns consistent with aggressive-tumour phenotypes, acquiring metastatic activity *in vivo*. SATB1 delineates specific epigenetic modifications at target gene loci, directly upregulating metastasis-associated genes while downregulating tumour-suppressor genes. SATB1 reprogrammes chromatin organisation and the expression profiles of breast tumours to promote growth and metastasis [107].

The upregulation found on Affymetrix GeneChip could not be confirmed by RT-qPCR (Figure 75).

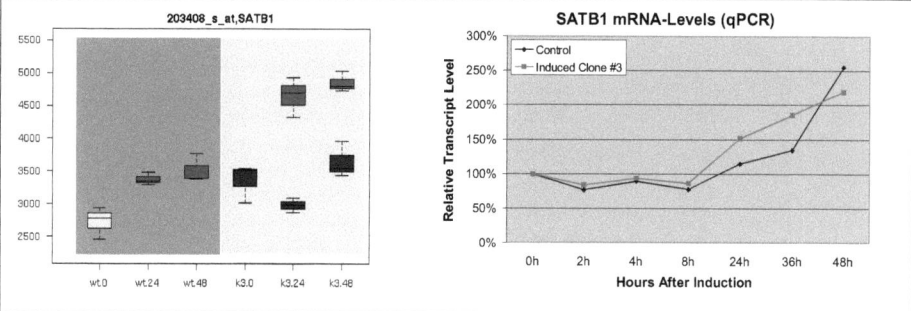

Figure 75. SATB1 mRNA levels in induced CLCA2-expressing T47D Tet-on clone #3 versus control
Left: Affymetrix microarray studies using Affymetrix GeneChip Human Genome U133 Array
The MAS5-normalised expression levels of SATB1 mRNA are shown in box-plots. The vertical centre line in the box indicates the median; the box itself represents the interquartile range (IQR) between the first and third quartiles. Whiskers extend to 1.5 times the IQR. In the left dark grey field, expression levels in control cells (T47D Tet-on) are shown. In the right light grey field, expression levels in T47D Tet-on clone #3 (k3) are represented.
Light grey box-plots indicate SATB1 transcript levels of cells with induced CLCA2 expression; dark grey box-plots represent SATB1 transcript levels of non-induced cells. The white box-plot represents the transcript level in control cells at time point of induction. 0, 24 and 48 indicate the time points (h) of lysate preparation post induction.
Right: RT-qPCR in duplex assay with β-2-microglobulin as internal reference for normalisation
Normalised SATB1 mRNA levels of induced clone #3 are shown in relation to mRNA levels in control cells (T47D Tet-on with empty expression vector).

EP300

The transcriptional coactivator E1A binding protein p300 is a ubiquitous nuclear phosphoprotein and transcriptional cofactor with intrinsic acetyltransferase activity. p300 controls the expression of numerous genes in cell-type and signal-specific manner, and plays a pivotal role in cellular proliferation, apoptosis, and embryogenesis [108].

Although the expression profile of EP300 was contradictory, it was analysed by RT-qPCR and found to be downregulated (Figure 76). One reason for this discrepancy might be the presence of various isoforms or splice variants of EP300, generating a contradictory profile on Affymetrix GeneChips.

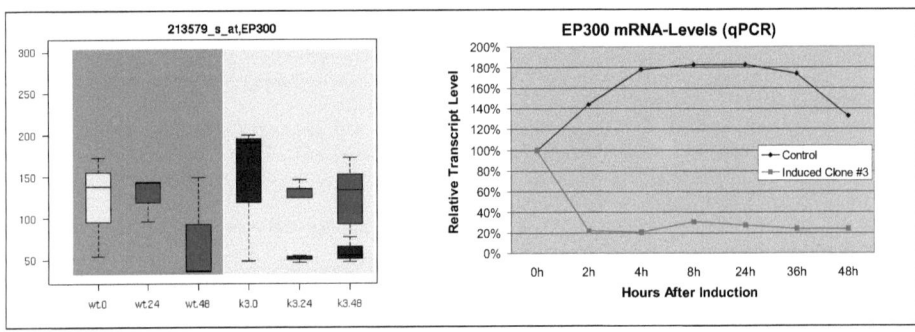

Figure 76. EP300 mRNA levels in induced CLCA2-expressing T47D Tet-on clone #3 versus control
Left: Affymetrix microarray studies using Affymetrix GeneChip Human Genome U133 Array
The MAS5-normalised expression levels of EP300 mRNA are shown in box-plots. The vertical centre line in the box indicates the median; the box itself represents the interquartile range (IQR) between the first and third quartiles. Whiskers extend to 1.5 times the IQR. In the left dark grey field, expression levels in control cells (T47D Tet-on) are shown. In the right light grey field, expression levels in T47D Tet-on clone #3 (k3) are represented.
Light grey box-plots indicate EP300 transcript levels of cells with induced CLCA2 expression; dark grey box-plots represent EP300 transcript levels of non-induced cells. The white box-plot represents the transcript level in control cells at the time point of induction. 0, 24 and 48 indicate the time points (h) of lysate preparation post induction.
Right: RT-qPCR in duplex assay with β-2-microglobulin as internal reference for normalisation
Normalised EP300 mRNA levels of induced clone #3 are shown in relation to mRNA levels in control cells (T47D Tet-on with empty expression vector).

LTBP1

Overexpression of Latent transforming growth factor beta binding protein 1 (LTBP1) in association with TGF-beta 1 was found in ovarian carcinoma [109].
LTBP1 is impressively upregulated when analysed on GeneChips. Although RT-qPCR confirmed upregulation, the level of upregulation is significantly lower (Figure 77).

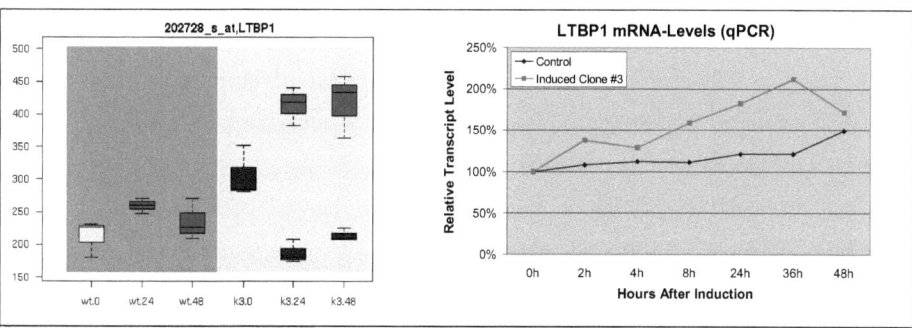

Figure 77. LTBP1 mRNA levels in induced CLCA2-expressing T47D Tet-on clone #3 versus control
Left: Affymetrix microarray studies using Affymetrix GeneChip Human Genome U133 Array
The MAS5-normalised expression levels of LTBP1 mRNA are shown in box-plots. The vertical centre line in the box indicates the median; the box itself represents the interquartile range (IQR) between the first and third quartiles. Whiskers extend to 1.5 times the IQR. In the left dark grey field, expression levels in control cells (T47D Tet-on) are shown. In the right light grey field, expression levels in T47D Tet-on clone #3 (k3) are represented.
Light grey box-plots indicate LTBP1 transcript levels of cells with induced CLCA2 expression; dark grey box-plots represent LTBP1 transcript levels of non-induced cells. The white box-plot represents the transcript level in control cells at the time point of induction. 0, 24 and 48 indicate the time points (h) of lysate preparation post induction.
Right: RT-qPCR in duple xassay with β-2-microglobulin as internal reference for normalisation
Normalised LTBP1 mRNA levels of induced clone #3 are shown in relation to mRNA levels in control cells (T47D Tet-on with empty expression vector).

TSC1

Tuberous sclerosis complex 1 is a tumour suppressor gene which is involved in the mTOR pathway. Interestingly, a recently published study suggests the involvement of TSC genes and other members of the mTOR signaling pathway in the pathogenesis of oral squamous cell carcinoma [110].

TSC1 was also shown to be involved in induction of autophagy. The role of autophagy in oncogenesis is manifold: At the early stages of tumor formation, autophagy acts as a tumor suppressor. During tumor progression, autophagy contributes to tumor growth. For instance, inhibition of autophagy by siRNAs targeting essential autophagy genes such as TSC1, sensitises cancer cells to the induction of apoptosis by radiotherapy and a wide range of chemotherapeutic agents. Inhibition of autophagy sensitises breast cancer cells to killing by the estrogen receptor antagonist tamoxifen, prostate cancer cells to androgen deprivation, colon cancer cells to amino-acid or glucose deprivation, and Bax-/- HCT116 cells to TRAIL-induced apoptosis. Similar effects have been observed sensitising tumor cells to anoikis (apoptosis due to detachment from the extracellular matrix) [111].

Upregulation of TSC1 in cells with induced CLCA2 expression could be confirmed by both methods (Figure 78).

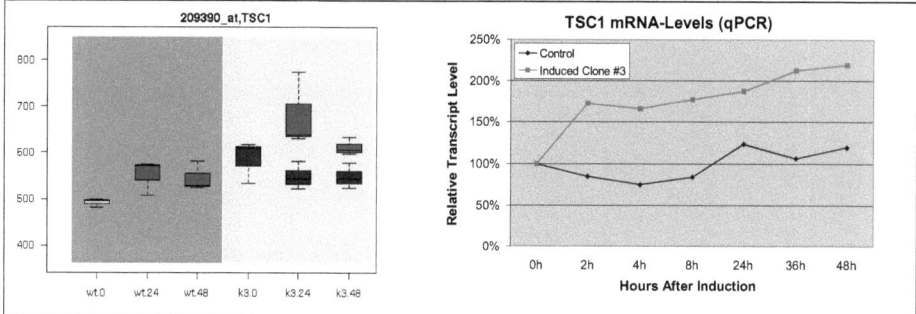

Figure 78. TSC1 mRNA levels in induced CLCA2-expressing T47D Tet-on clone #3 versus control
Left: Affymetrix microarray studies using Affymetrix GeneChip Human Genome U133 Array
The MAS5-normalised expression levels of TSC1 mRNA are shown in box-plots. The vertical centre line in the box indicates the median; the box itself represents the interquartile range (IQR) between the first and third quartiles. Whiskers extend to 1.5 times the IQR. In the left dark grey field, expression levels in control cells (T47D Tet-on) are shown. In the right light grey field, expression levels in T47D Tet-on clone #3 (k3) are represented.
Light grey box-plots indicate TSC1 transcript levels of cells with induced CLCA2 expression; dark grey box-plots represent TSC1 transcript levels of non-induced cells. The white box-plot represents the transcript level in control cells at the time point of induction. 0, 24 and 48 indicate the time points (h) of lysate preparation post induction.
Right: RT-qPCR in duplex assay with β-2-microglobulin as internal reference for normalisation
Normalised TSC1 mRNA levels of induced clone #3 are shown in relation to mRNA levels in control cells (T47D Tet-on with empty expression vector).

MACF1

Microtuble-actin crosslinking factor 1 is involved in the Wnt signaling pathway [112]. An upregulation after CLCA2-induction could be observed on the Affymetrix GeneChip and at later time points applying RT-qPCR (Figure 79).

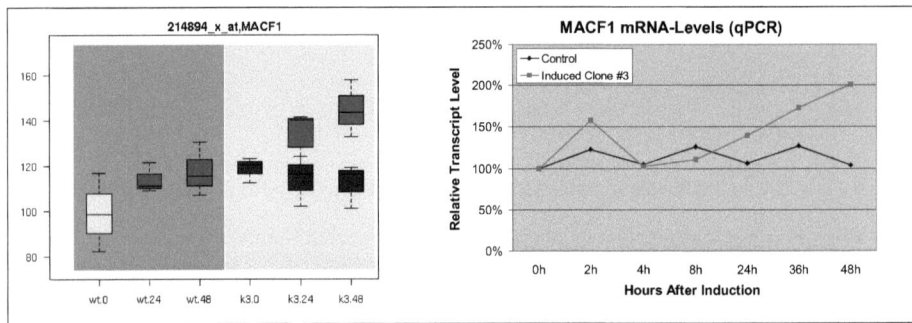

Figure 79. MACF1 mRNA levels in induced CLCA2-expressing T47D Tet-on clone #3 versus control
Left: Affymetrix microarray studies using Affymetrix GeneChip Human Genome U133 Array
The MAS5-normalised expression levels of MACF1 mRNA are shown in box-plots. The vertical centre line in the box indicates the median; the box itself represents the interquartile range (IQR) between the first and third quartiles. Whiskers extend to 1.5 times the IQR. In the left dark grey field, expression levels in control cells (T47D Tet-on) are shown. In the right light grey field, expression levels in T47D Tet-on clone #3 (k3) are represented.
Light grey box-plots indicate MACF1 transcript levels of cells with induced CLCA2 expression; dark grey box-plots represent MACF1 transcript levels of non-induced cells. The white box-plot represents the transcript level in control cells at the time point of induction. 0, 24 and 48 indicate the time points (h) of lysate preparation post induction.
Right: RT-qPCR in duplex assay with β-2-microglobulin as internal reference for normalisation
Normalised MACF1 mRNA levels of induced clone #3 are shown in relation to mRNA levels in control cells (T47D Tet-on with empty expression vector).

CHD6

Chromodomain helicase DNA binding protein 6 is a DNA-dependent ATPase that localises at nuclear sites of mRNA synthesis [113]. CHD proteins have drawn increased attention, because some of them were found to form large multi-subunit complexes, involved in transcription-related events like gene activation, suppression, or histone modification [113].

CHD6 was found upregulated after CLCA2-induction both, on the Affymetrix GeneChips and by RT-qPCR (Figure 80).

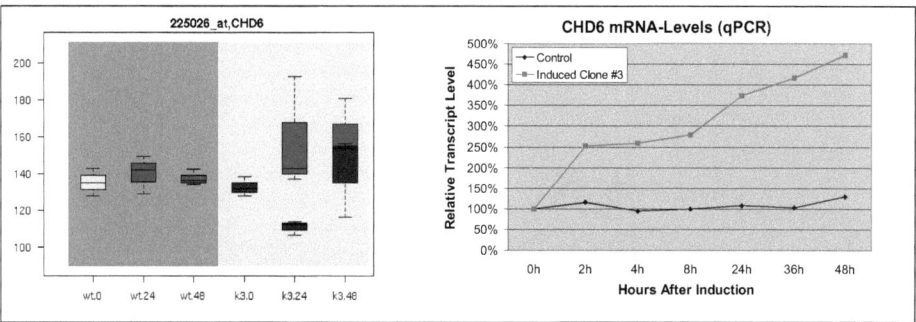

Figure 80. CHD6 mRNA levels in induced CLCA2-expressing T47D Tet-on clone #3 versus control
Left: Affymetrix microarray studies using Affymetrix GeneChip Human Genome U133 Array
The MAS5-normalised expression levels of CHD6 mRNA are shown in box-plots. The vertical centre line in the box indicates the median; the box itself represents the interquartile range (IQR) between the first and third quartiles. Whiskers extend to 1.5 times the IQR. In the left dark grey field, expression levels in control cells (T47D Tet-on) are shown. In the right light grey field, expression levels in T47D Tet-on clone #3 (k3) are represented.
Light grey box-plots indicate CHD6 transcript levels of cells with induced CLCA2 expression; dark grey box-plots represent CHD6 transcript levels of non-induced cells. The white box-plot represents the transcript level in control cells at the time point of induction. 0, 24 and 48 indicate the time points (h) of lysate preparation post induction.
Right: RT-qPCR in duplex assay with β-2-microglobulin as internal reference for normalisation
Normalised CHD6 mRNA levels of induced clone #3 are shown in relation to mRNA levels in control cells (T47D Tet-on with empty expression vector).

RUTBC3

RUN and TBC1 domain containing 3 gene belongs to the RUN domain protein family [114]. RUN domains are present in several proteins that are linked particularly to the functions of GTPases in the Rap and Rab families. They could hence play an important role in multiple Ras-like GTPase signaling pathways [115]. RUTBC3 was found to be upregulated after induction of CLCA2 expression in both assays (Figure 81).

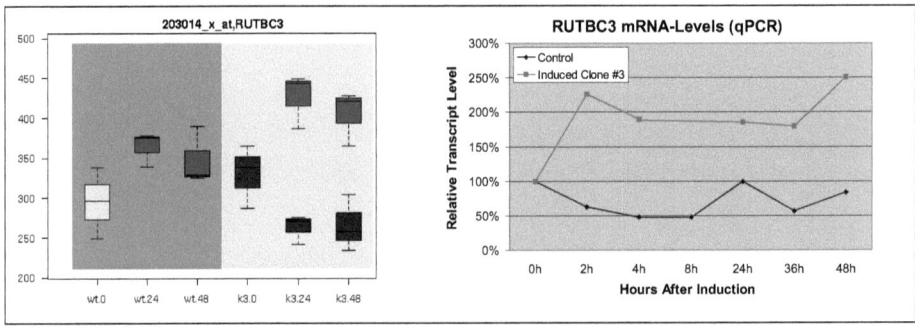

Figure 81. RUTBC3 mRNA levels in induced CLCA2-expressing T47D Tet-on clone #3 versus control
Left: Affymetrix microarray studies using Affymetrix GeneChip Human Genome U133 Array
The MAS5-normalised expression levels of RUTBC3 mRNA are shown in box-plots. The vertical centre line in the box indicates the median; the box itself represents the interquartile range (IQR) between the first and third quartiles. Whiskers extend to 1.5 times the IQR. In the left dark grey field, expression levels in control cells (T47D Tet-on) are shown. In the right light grey field, expression levels in T47D Tet-on clone #3 (k3) are represented.
Light grey box-plots indicate RUTBC3 transcript levels of cells with induced CLCA2 expression; dark grey box-plots represent RUTBC3 transcript levels of non-induced cells. The white box-plot represents the transcript level in control cells at the time point of induction. 0, 24 and 48 indicate the time points (h) of lysate preparation post induction.
Right: RT-qPCR in duplex assay with β-2-microglobulin as internal reference for normalisation
Normalised RUTBC3 mRNA levels of induced clone #3 are shown in relation to mRNA levels in control cells (T47D Tet-on with empty expression vector).

P4HA2

Prolyl-4 hydroxylase-2 belonging to the procollagen hydroxylase protein family was shown earlier to be differentially regulated upon ERBB2 (HER2/neu) overexpression in human mammary luminal epithelial cells [116]. Furthermore, P4HA2 is suggested to be a potent molecular classifier able to discriminate between papillary thyroid carcinoma and non-malignant thyroid [117]. In this study P4HA2 was found to be upregulated after CLCA2-induction when analysed on Affymetrix GeneChips. However, no specific CLCA2-driven induction of this gene was detected in the RT-qPCR (Figure 82).

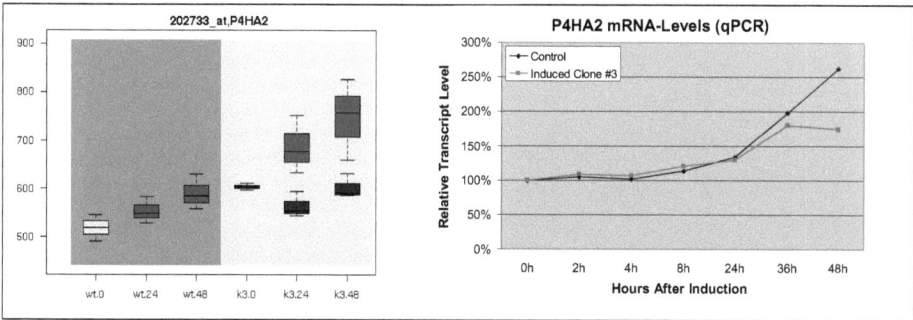

Figure 82. P4HA2 mRNA levels in induced CLCA2-expressing T47D Tet-on clone #3 versus control
Left: Affymetrix microarray studies using Affymetrix GeneChip Human Genome U133 Array
The MAS5-normalised expression levels of P4HA2 mRNA are shown in box-plots. The vertical centre line in the box indicates the median; the box itself represents the interquartile range (IQR) between the first and third quartiles. Whiskers extend to 1.5 times the IQR. In the left dark grey field, expression levels in control cells (T47D Tet-on) are shown. In the right light grey field, expression levels in T47D Tet-on clone #3 (k3) are represented.
Light grey box-plots indicate P4HA2 transcript levels of cells with induced CLCA2 expression; dark grey box-plots represent P4HA2 transcript levels of non-induced cells. The white box-plot represents the transcript level in control cells at the time point of induction. 0, 24 and 48 indicate the time points (h) of lysate preparation post induction.
Right: RT-qPCR in duplex assay with β-2-microglobulin as internal reference for normalisation
Normalised P4HA2 mRNA levels of induced clone #3 are shown in relation to mRNA levels in control cells (T47D Tet-on with empty expression vector).

ASPM

Asp (abnormal spindle)-like, microcephaly associated (drosophila) gene was recently found as a novel marker for vascular invasion, early recurrence, and poor prognosis of hepatocellular carcinoma [118]. Overexpression after CLCA2-induction was found on the Affymetrix GeneChip. Although a similar expression profile in CLCA2-induced and control cells was shown by RT-qPCR, the levels of CLCA2 transcript were higher in CLCA2-expressing cells (Figure 83).

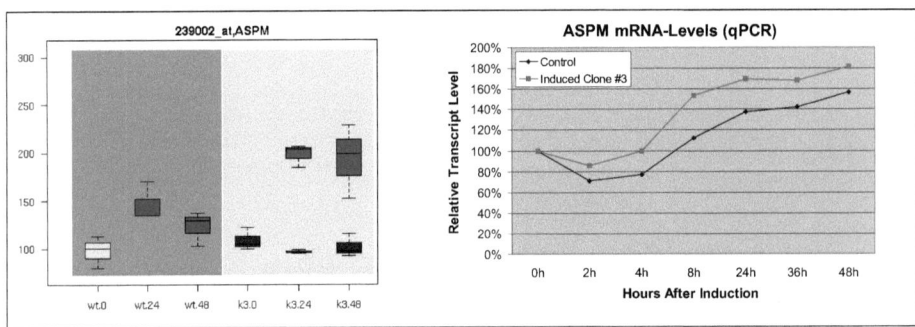

Figure 83. ASPM mRNA levels in induced CLCA2-expressing T47D Tet-on clone #3 versus control
Left: Affymetrix microarray studies using Affymetrix GeneChip Human Genome U133 Array
The MAS5-normalised expression levels of ASPM mRNA are shown in box-plots. The vertical centre line in the box indicates the median; the box itself represents the interquartile range (IQR) between the first and third quartiles. Whiskers extend to 1.5 times the IQR. In the left dark grey field, expression levels in control cells (T47D Tet-on) are shown. In the right light grey field, expression levels in T47D Tet-on clone #3 (k3) are represented.
Light grey box-plots indicate ASPM transcript levels of cells with induced CLCA2 expression; dark grey box-plots represent ASPM transcript levels of non-induced cells. The white box-plot represents the transcript level in control cells at the time point of induction. 0, 24 and 48 indicate the time points (h) of lysate preparation post induction.
Right: RT-qPCR in duplex assay with β-2-microglobulin as internal reference for normalisation
Normalised ASPM mRNA levels of induced clone #3 are shown in relation to mRNA levels in control cells (T47D Tet-on with empty expression vector).

RESULTS

To identify CLCA2 coregulated genes upon induced CLCA2 expression, expression profiles at different time points were analysed and mainly confirmed by RT-qPCR. These analyses were performed with two limitations: i) only a small number of genes out of a about 250 statistically relevant and differentially induced genes were selected for detailed analysis and ii) both assays (hybridisation on GeneChips and RT-qPCR) were performed only once, but samples from different time points were analysed. Nevertheless, as seen in Table 3 a good correlation between both approaches was achieved, only 2 out of 18 analysed expression profiles are questionable (Table 3).

GENE ID	Chip	qPCR
AFF4	u	u
ASPM	u	u?
C10orf118	u	u
CDIPT	u	u
CHD6	u	u
CLCA2	u	u
DTX2	u	u
EP300	d/i	d
LTBP1	u	u
MACF1	u	u
MLL2	u	u
P4HA2	i	i
RUTBC3	u	u
SATB1	i	i
SFI1	u	u
TP53BP1	i	i
TSC1	u	u
UNC84A	u	u

u.....upregulated
d.....downregulated
i......indifferent

Table 3. Comparison of expression data derived from Affymetrix microarray analysis and RT-qPCR
Status of gene regulation from microarray and RT-qPCR studies are opposed.

Linking Profiles of Cells with Induced CLCA2 Expression to Canonical Pathways

As described above, a comprehensive study was performed on Affymetrix GeneChips in clones with inducible CLCA2 expression. Approximately 150 genes were shown to be specifically up- or downregulated in a statistically relevant mode (fold change > 1.8; p-value < 0.0001). In order to link these expression profiles of regulated genes to canonical pathways and bring them into a biological context, these data were analysed with the Ingenuity Pathways Analysis (IPA; Software of Ingenuity Systems Inc.). This program is a text- and data-mining program, which not only links gene and protein functions/interactions based on the knowledge of published data in the most relevant journals, but also allows to bring these interactions into a context of biological functions, regulatory cascades and functional groups (e.g. oncology/metastasis/prostate). It has to be mentioned that the presented studies can only be regarded as preliminary as the outcome of these *in silico* analyses have to be proven in experimental settings. On the other hand these data allow a much more focussed procedure for future experiments as this program not only describes protein-protein interactions, but also other types of reciprocal action such as protein-nucleic acid interactions. Based on these *in silico* analyses several interaction-networks were identified. Some exemplary statistically relevant networks are shown in Figure 85, Figure 86, Figure 87, and Figure 88.

For understanding the display of the following networks, an overview of the used symbols is given in Figure 84.

RESULTS

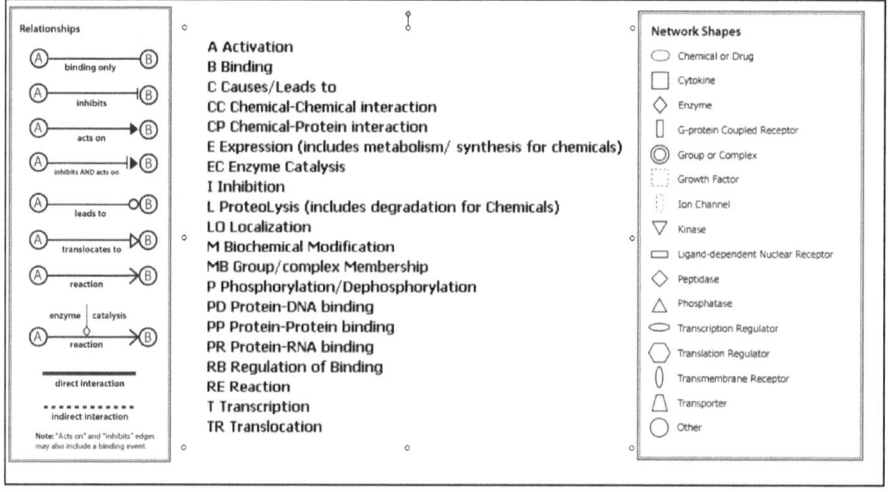

Figure 84. Abbreviations and explanation of relationships and network shapes used for Ingenuity Pathway Analyst (IPA) Software of Ingenuity Systems, Inc.

RESULTS

Network I (Figure 85) indicates that a group of specifically induced genes upon induction of CLCA2 is strongly linked to the interaction-axis of SMAD3–TP53–MYC.

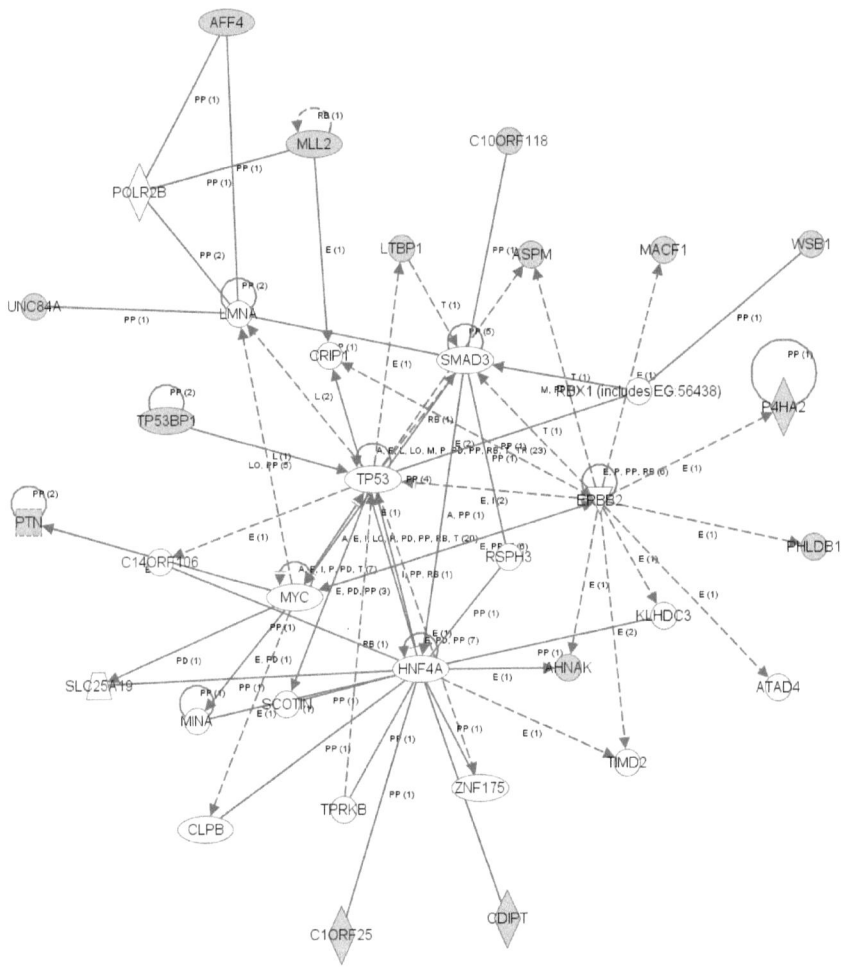

Figure 85. Network I: Link to the interaction-axis of SMAD3–TP53–MYC
Display of interactions based on the Ingenuity Pathway Analyst (IPA) Software of Ingenuity Systems, Inc.
Grey-coloured genes have been identified on Affymetrix GeneChips as differentially expressed upon CLCA2 overexpression. A detailed description of symbols and abbreviations used is given in Figure 84.

Network II (Figure 86) indicates a link to the TNF–IL2–IL10 interaction network. Interestingly, it has been shown recently that treatment of SCCs of the skin with a Toll-like receptor agonist such as imiquimod lead to a decrease in regulatory T (T reg) cells, producing less IL10 and TGF-beta, thereby inhibiting their suppressive activity. This means that SCCs evade the immune response at least in part by recruiting T reg cells *via* modulation of the IL10 pathway [119]. Additionally, in this network also CLCA2 is interacted *via* an indirect mode of action.

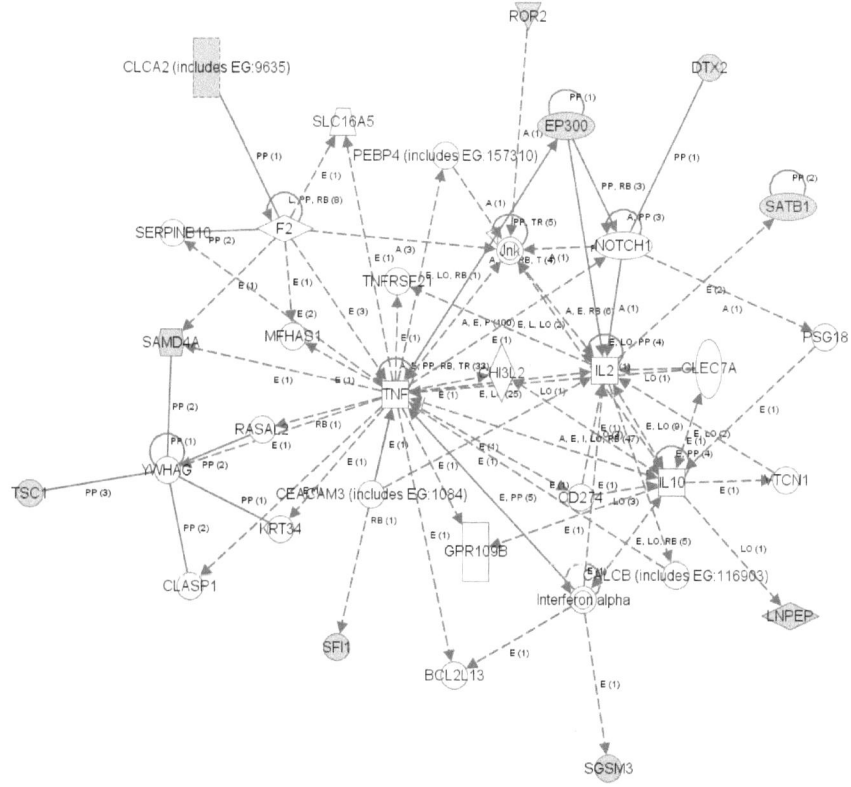

Figure 86. Network II: Link to the TNF–IL2–IL10 interaction network
Display of interactions based on the Ingenuity Pathway Analyst (IPA) Software of Ingenuity Systems, Inc.
Grey-coloured genes have been identified on Affymetrix GeneChips as differentially expressed upon CLCA2 overexpression. A detailed description of symbols and abbreviations used is given Figure 84.

Network III (Figure 87) brings some of the differentially upregulated genes into the context of the ERK–MAPK signaling cascade with ERK as the knot of this network.

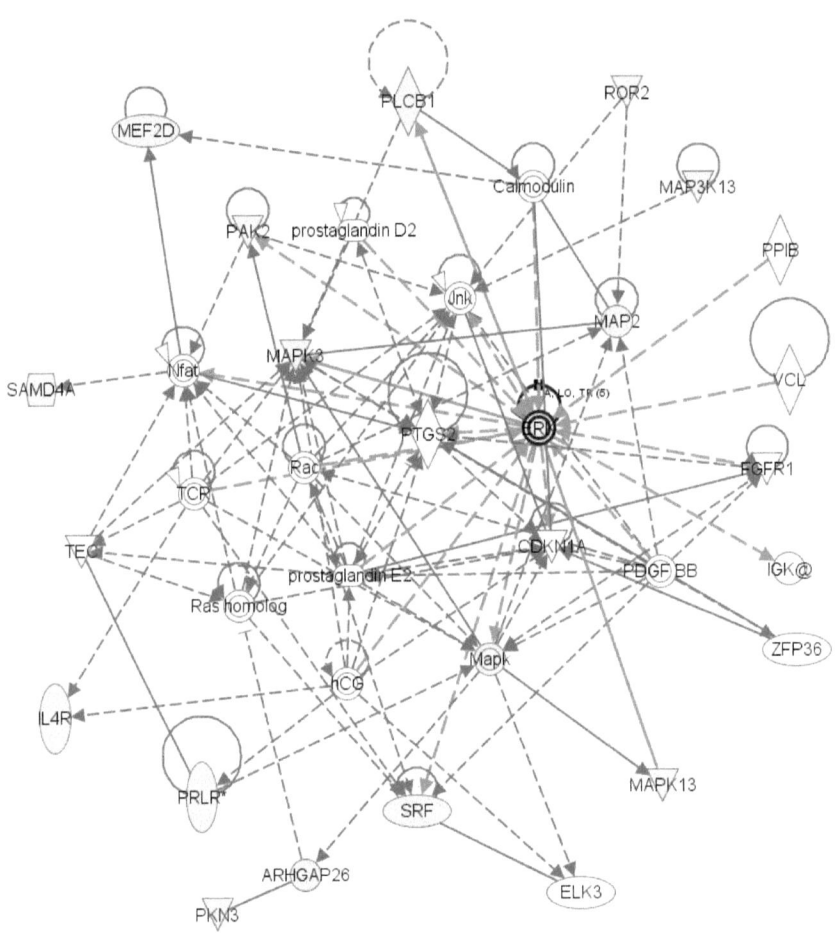

Figure 87. Network III: Link to the ERK–MAPK pathway
Display of interactions based on the Ingenuity Pathway Analyst (IPA) Software of Ingenuity Systems, Inc.
Red-coloured genes have been identified on Affymetrix GeneChips as differentially expressed upon CLCA2 overexpression. A detailed description of symbols and abbreviations used is given in Figure 84. Lines/arrows in blue indicate interactions of ERK.

RESULTS

In Figure 88 merging of the most significant networks is shown, focussing excusively on protein-protein interactions. As already mentioned, utilising this program, also other interactions such as protein-DNA interactions can be visualised. For the sake of simplicity only the protein-protein interaction is presented, highlightening biologically interesting clusters such as the cluster of v-src (involved in regulation of many different signaling cascades) or the polypyrimidine tract binding protein 1 (PTBP1) which is essentially involved in translational control (Figure 88).

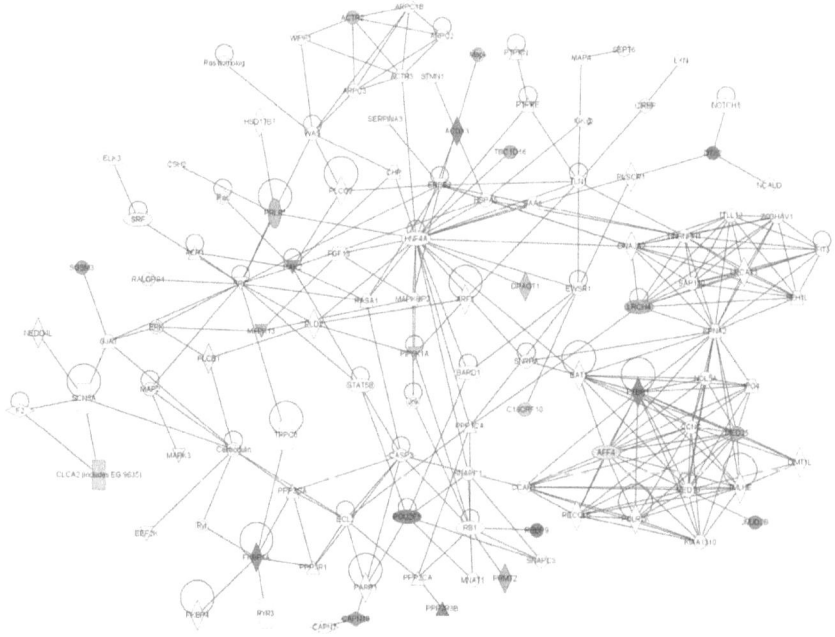

Figure 88. Merging most significant networks
Display of interactions based on the Ingenuity Pathway Analyst (IPA) Software of Ingenuity Systems, Inc.
Red-coloured genes have been identified on Affymetrix GeneChips as differentially expressed upon CLCA2 overexpression. The intensity of the red colour indicates the level of differential regulation on GeneChips. A detailed description of symbols and abbreviations used is given in Figure 84.

Phospho-Proteomic Studies: Canonical Signaling Transduction Pathways

Utilising the IPA software (see above), only data derived from expression profiling were applied so far. In order to learn more about the contribution of some important executer proteins of canonical pathways, which might be found activated/deactivated (=phosphorylated/dephosphorylated) in the experimental setting of CLCA2 induction, cell lysates which were taken at various time points of induced and non-induced T47D Tet-on clone #3 cells, were analysed on Western blots and on phospho-proteomic antibody arrays.

RESULTS

Western Blots

Antibodies listed in Table 4 were used for Western blots.

Target Protein Name	Phospho Site	Company	Cat. No.
Akt	S473	Cell Signaling	#9271
Akt	S308	Cell Signaling	#9275
Akt	Y326	Cell Signaling	#2968
Akt	pan	Cell Signaling	#9272
GSK-3beta	S9	Cell Signaling	#9336
PTEN	S380	Cell Signaling	#9551
PDK1	S241	Cell Signaling	#3061
PDK1	pan	Cell Signaling	#3062
PI3K	p85 Y458 / p55 Y199	Cell Signaling	#4228
FAK	Y861	Biosource	44-626G
iKKalpha/beta	S176/180	Cell Signaling	#2697S
NF-KB p65	S536	Cell Signaling	#3031
NF-KB p65	pan	Cell Signaling	#3034
MEK 1/2	S217/221	Cell Signaling	#9121
p44/42 MAP Kinase	T202/Y204	Cell Signaling	#9101
p38 MAPK	T180/Y182	Cell Signaling	#9211
p90RSK	S380	Cell Signaling	#9341
RSK1	S221/S227	Biosource	44-924G
RSK1	S363/S369	Biosource	44-926G
Elk1	S383	Cell Signaling	#9181
Raf	S259	Cell Signaling	#9421
c-Raf	S338	Biosource	#9427
ERK 1 & 2	T185/Y187	Biosource	44-680G
Erk 1&2	pan	Biosource	44-654G
p70 S6 Kinase	T389	Cell Signaling	#9234S
p70 S6 Kinase	pan	Cell Signaling	#9202
mTOR	S2448	Cell Signaling	#2971
mTOR	S2481	Cell Signaling	#2974
4E-BP1	T37/46	Cell Signaling	#2855
4E-BP1	T70	Cell Signaling	#9455
4E-BP1	S65	Cell Signaling	#9451
FoxO1	S319	Cell Signaling	#2487
FoxO3a	S318/321	Cell Signaling	#9465
FoxO3a	S253	Cell Signaling	#9466
Stat1	Y701	Cell Signaling	#9171
Stat2	Y690	Cell Signaling	#4471
Stat3	Y705	Cell Signaling	#9131
Stat3	S727	Cell Signaling	#9134
Stat3	Y705	Cell Signaling	#9131
Stat5	Y694	Cell Signaling	#9351
Stat6	Y641	Cell Signaling	#9361
NPM/B23 (Nucleophosmin 1)	T234/T237	BioLegend	#619101
PKCe	pan	Biosource	AHO0743
p27	pan	novocastra	NCL-p27
Histon H3	S10	Upstate	#06-570
TSC2	T1462	Cell Signaling	#3611
TSC2	S1254	Cell Signaling	#3616
KI67	pan	Dako Cytomation	M7240

phosphorylated
phosphorylated ?

dephosphorylated
dephosphorylated ?

Table 4. Phospho-antibodies tested on protein lysates of induced/non-induced CLCA2-expressing T47D Tet-on clone #3
Genes marked in green were found to be phosphorylated upon induction of CLCA2 expression.
Genes marked in yellow were found to be dephosphorylated upon induction of CLCA2 expression.

RESULTS

In Figure 89 Western blots of proteins which were found to be phopshorylated upon CLCA2-induction are shown. An increase in phosphorylation of AKT (S473; Figure 89a), ERK 1&2 (T185/Y187; Figure 89c), MAPK (T202/Y204; Figure 89d), RAF (S259; Figure 89e), CRAF (S338; Figure 89f), and PDK1 (S241; Figure 89g) was observed. No change in the phosphorylation pattern was seen for AKT at position S308. These data confirm previous predictions made through the usage of the IPA software, indicating a link to the ERK–MAPK canonical pathway.

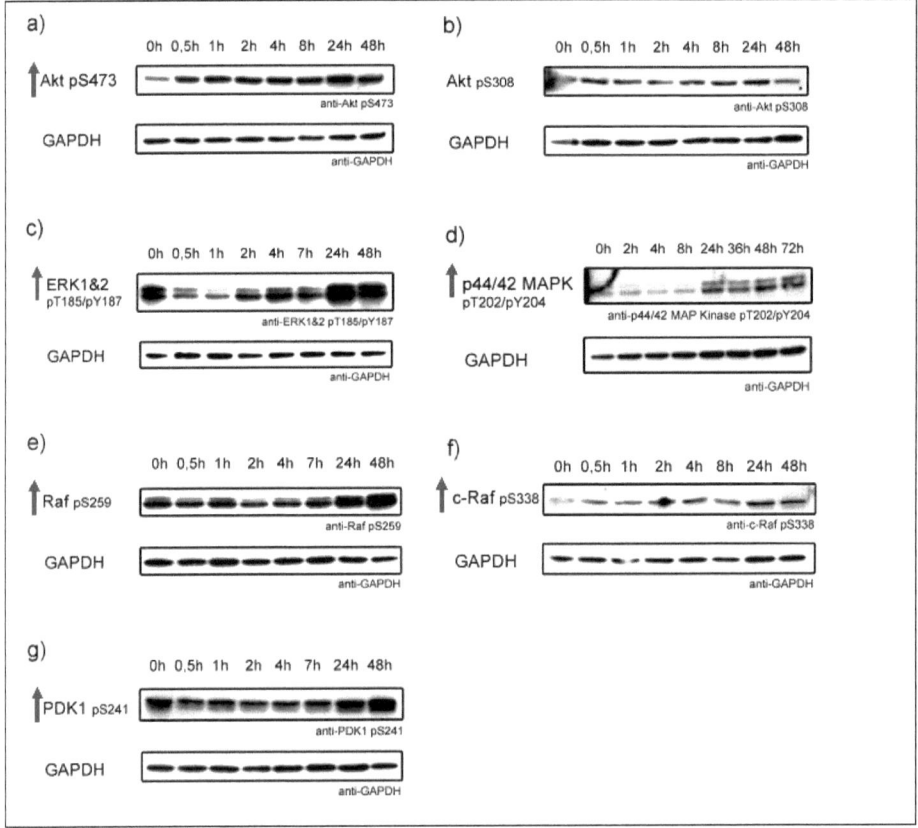

Figure 89. Proteins involved in the AKT–ERK–MEK canonical pathway exhibiting an increase in phosphorylation upon induction of CLCA2 expression
Expression of CLCA2 was induced in CLCA2-expressing T47D Tet-on clone #3 by addition of Doxycycline. Protein lysates were harvested at indicated time points post induction. CLCA2 expression of lysates was confirmed before use. For Western blot 50 μg protein was loaded per lane; the phosphorylation status of indicated proteins was tested with a respectiv phospho-antibody (for detailed information on antibodies see Table 4). GAPDH served as loading control.

In Figure 90 Western blots of identical protein lysates are shown. In contrast to Figure 83, proteins analysed were found dephosphorylated upon CLCA2-induction including mTOR (p2448; Figure 90a), 4E-BP1 (T37/T47; Figure 90b), STAT1 (Y701; Figure 90c), and NPM (T234/T237; Figure 90d), respectively. Nucleophosmin (NPM), for instance, is involved in cytoskeletal regulation [120], whereas deregulation of STAT1 indicates a link of CLCA2 expression to translational control.

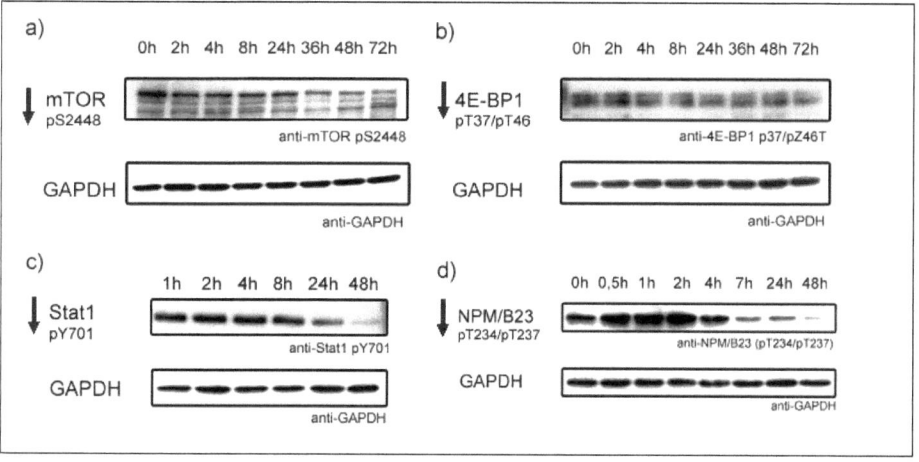

Figure 90. Proteins exhibiting a decrease in phosphorylation upon induction of CLCA2 expression
Expression of CLCA2 was induced in CLCA2-expressing T47D Tet-on clone #3 by addition of Doxycycline.
Protein lysates were harvested at indicated time points post induction. CLCA2 expression of lysates was confirmed before use. For Western blot 50 µg protein was loaded per lane; the phosphorylation status of indicated proteins was tested with a respectiv phospho-antibody (for detailed information on antibodies see Table 4). GAPDH served as loading control.

Western blots of those antibodies listed in Table 4 which did not show any effect on the phosphorylation status of the respective protein upon CLCA2-induction, are not shown.

RESULTS

Antibody Arrays: Proteome Profiler

"Human Phospho-MAPK Array Kit" (Cat.# ARY002) and "Human Phospho-RTK Array Kit" (Cat.# ARY001) of R&D Systems were tested on lysates of CLCA2-expressing T47D Tet-on clones and on CLCA2-constitutively-expressing T47D clone.

The principle of the assays is that capture and control antibodies have been spotted in duplicates on nitrocellulose membranes. Cell lysates are diluted and incubated with the "Human Phospho-Array Kit". After binding both, phosphorylated and non-phosphorylated proteins, unbound material is washed off. A cocktail of phospho-site-specific biotinylated antibodies is then used to detect phosphorylated kinases via Streptavidin-HRP and chemiluminescence.

The phosphorylation status of all three major families of mitogen-activated protein kinases (MAPKs), the extracellular signal-regulated kinases (ERK1/2), c-Jun N-terminal kinases (JNK1-3), different p38 isoforms ($\alpha/\beta/\delta/\gamma$), other intracellular kinases such as Akt, GSK-3, and p70 S6 kinases was analysed, using the "Human Phospho-MAPK Array Kit" from R&D Systems and for analysis of the relative levels of phosphorylation of 42 different receptor tyrosine kinases, the "Human Phospho-RTK Array Kit" from R&D Systems (Cat.# ARY001) was used.

As shown in Figure 91, the duplicates of B3 and B4, representing ERK1 and the duplicates C3 and C4, representing ERK3 (red squares) show a higher phosphorylation level in CLCA2-constitutively-expressing T47D clone #47, than in contol cells with the empty vector (blue squares).

CLCA2-constitutively-expressing T47D clone #47

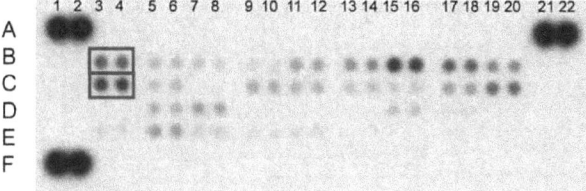

Control cells (empty vector)

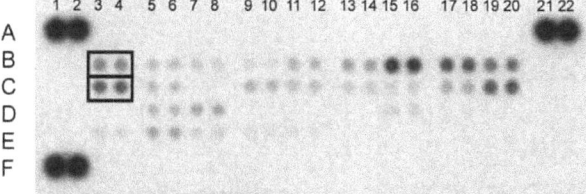

Figure 91. Phosphorylation status of MAPKs, ERK1/2, JNK1-3, different p38 isoforms and other intracellular kinases upon expression of CLCA2 on antibody arrays
Protein lysates of CLCA2-constitutively-expressing T47D Clone #47 and "pseudo-induced" control cells with empty vector were harvested; CLCA2 expression of lysates was confirmed before use. The lysates were tested for the phosphorylation status of various proteins of the MAPK pathway using the "Human Phospho-MAPK Array Kit" from R&D Systems (Cat.# ARY002).

As shown in Figure 92 the duplicates of B3 and B4, representing EGF-R and the duplicates B17 and B18, representing Insulin-R and the duplicated B19 and B20, representing IGF-I R (red squares) show a higher phosphorylation level in CLCA2-inducibly-expressing T47D clone #3, than in contol cells with the empty vector (blue squares). These data point to a possible activation of these three RTKs by CLCA2.

CLCA2-expressing T47D Tet-on clone #3

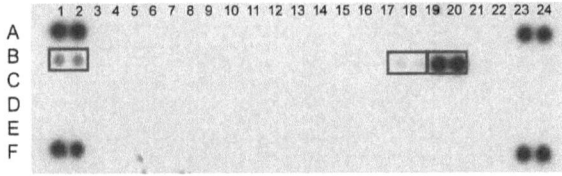

Control cells (empty vector)

Figure 92. Phosphorylation status of different receptor tyrosine kinases upon induction of CLCA2 expression on antibody arrays
Protein lysates of induced CLCA2-expressing T47D Tet-on Clone #3 and "pseudo-induced" control cells with the empty vector were harvested 48 h post induction; CLCA2 expression of lysates was confirmed before use.
The lysates were tested for the phosphorylation status of 42 different receptor tyrosine kinases using the "Human Phospho-RTK Array Kit" from R&D Systems (Cat.# ARY001).

DISCUSSION

Introductory Remarks: General Reflections on Targeted Therapy in Oncology

Cancer incidence is increasing worldwide with a tendency to be concentrated in developing countries (WHO estimation: 66 % of all new cases will be diagnosed within such regions). As the majority of patients are diagnosed in advanced stage, "classical" chemotherapy very often is still the first choice in treatment modality. As this therapy acts mainly in a non-specific manner, significant toxicity is observed. Indisputably, there is a high medical need to identify novel target molecules for more focused therapies in oncology. However, identification of such novel molecular targets requires a better understanding of the biology of cancer cells and their metabolic functioning. The knowledge about molecules and processes that are unique to cancerous cells or at least to a given cancer type is a prerequisite in the development of novel therapeutic approaches in cancer patients, leading to "targeted therapy".

Over the last few years, a set of potential targets for cancer-directed therapy has been explored. Targeted therapy is defined as a drug or molecule causing tumour cell kill by interacting with predefined target(s) present on malignant cells [121]. Besides others, such an approach has the obvious advantage that it would selectively attack molecules and/or signaling transduction cascades that are unique to the tumour cells. Such kind of therapy has the potential to exclude normal cells to be hit, resulting in the reduction of side effects and general toxicity of the therapy, to have greater efficacy and to improve quality of life. Depending on the mode of action and the defined targets, various agents can be classified in subcategories such as monoclonal antibodies like Tratuzimab [122], tyrosine kinase inhibitors [123-127], antiangiogenic agents [128;129], cyclin-dependent kinases (CDK) and mitotic kinase inhibitors [130-132], proteasome inhibitors [133], and Cox-2 inhibitors [134] to name but a few.

Targeted therapies generate additional benefits and a higher efficacy in tumour patients. However, both, benefits and efficacy need to be evaluated carefully [121].

DISCUSSION

Cancer is a multistep process and genetic events (deletions, amplifications, rearrangements, mutations, epigenetic regulation etc.) result in the activation of (proto-)oncogenes or the inactivation of tumour suppressor genes. As a result, the proliferation and growth of a tumour cell increases. It is of interest that the experimental inactivation of even a single oncogene can be sufficient to induce sustained tumour regression or vice versa, although tumourigenesis is thought to be a multistep process. Tumours become irrevocably addicted to these oncogenes. "Oncogene addiction" describes therefore, the phenomenon by which some cancers that contain multiple genetic and epigenetic abnormalities remain dependent on (addicted to) one or a few genes for both maintenance of the malignant phenotype and cell survival/proliferation. One explanation for this phenomenon is that activated oncogenes (e.g. mutated EGFR) result in a constitutive activation of a signaling cascade leading to sustained proliferation and survival [28]. On the other hand, a single oncogene is not sufficient to fulfill all requirements of a "full-blown" tumour cell and to overcome all physiologic barriers and safety mechanisms [16;135]. Therefore, it was suggested that oncogene activation initiates tumourigenesis mainly because it directly overrides physiologic programs inducing a state of "cellular amnesia", not only inducing constitutive proliferation, but also bypassing checkpoint mechanisms that are essential for cellular mortality, self-renewal, and genomic integrity [136;137]. Correspondingly, in a tumour exhibiting the status of "oncogene addiction", the inactivation of a single oncogene can restore some of these pathways resulting in proliferative arrest, differentiation, cellular senescence, and/or apoptosis. According to Felsher and colleagues "oncogenes induce cancer because they induce a cellular state of enforced oncogenic amnesia in which, only upon oncogene inactivation, the tumour becomes aware of its transgression" [138]. Novel techniques of systems biology will provide methods for analysing the entire circuitry of cancer cells and thus facilitate identification of these pathways of "oncogene addiction" or "oncogene amnesia" in specific types of human cancer. Regardless whether we call it "oncogene addiction" or "oncogene amnesia", a prerequisite for the development of a targeted therapy is the identification of tumour-specific target genes which might not only be used as targets for mono-therapies but also be applied in combination with other anticancer agents [139].

Oncogenes and tumour suppressors are derived from all protein families, not only bearing an enzymatic function such as kinases, GPCRs, phosphatases, oxidoreductases, proteases etc., but also exhibiting transport [140;141], transcriptional [142;143] or structural/topologic functions [144;145]. For many of these proteins the contribution to carcinogenesis has been univocally demonstrated (e.g. receptor tyrosine kinases), for others it remains to be determined (e.g. transporters, ion channels).

Often tumour-associated overexpression is the first hint that a gene might be involved in tumourigenesis. However, overexpression per se explains neither its essential contribution to tumour growth, survival or maintenance nor whether a gene (gene function) is able to drive tumour formation ("oncogene addiction"). A expression profile also does not allow to draw any conclusions about the mode of action of how a certain gene becomes overexpressed (e.g. chromosomal amplification, epigenetic regulation).

In the context of targeted therapy, ion channels are representing promising players. Specific stages of cancer progression are marked and regulated by expression and activity of different channels. Their contribution to the neoplastic phenotype ranges from control of cell proliferation and apoptosis, to regulation of invasiveness and metastatic spread. Often the contribution of these channels are independent of ion flow (e.g. shown for potassium (K^+) channels) but the underlying molecular mechanisms are poorly understood [146]. Ion channels frequently exert pleiotropic effects on the tumour cell such as regulating membrane potential and thereby controlling calcium ion (Ca^{2+}) fluxes and subsequently interfering with the cell cycle machinery [146;147]. The effect on mitosis might depend on the regulation of cell volume as shown in HeLa cells [146;148]. However, ion channels seem also to play an important role at late stage tumourigenesis by supporting angiogenesis, mediating the cell-matrix interaction, and regulating cell motility e.g. AQP1-null mice show defective tumour angiogenesis resulting from impaired endothelial cell migration whereas AQP3-null mice are resistant to skin tumourigenesis [140]. This is of interest as during this study it was found that CLCA2 might play an important role during colonisation of mycosis fungoides (cutaneous T cell lymphoma) into the skin (see section "*In Silico* Analysis of Expression Data").

Another important contribution of ion channels to tumourigenesis is executed via formation of protein complexes with other membrane proteins such as integrins or growth factor receptors. These interactions may trigger a variety of different signaling cascades. First evidence that calcium channel blockers suppress the growth of human cancer cells, both *in vitro* and *in vivo* has been published recently [149]. A major drawback of all these approaches is that often ion channel blockers produce serious side effects, such as cardiac arrhythmias and are not yet sufficiently specific enough [146]. This also holds true for members of the CLCA protein family for which no specific blocking compounds are available yet.

In Silico Analysis of Expression Data

Prior to the biological characterisation of CLCA2 a detailed *in silico* expression profile analysis was performed to learn more about its contribution to tumourigenesis. Based on the PhD-Thesis of Ulrich König and the Diploma Thesis of Stefan Amatschek [1], a lung SCC-specific expression profile of the human calcium-activated chloride channel, family member 2 (CLCA2) was suggested. Therefore, it was started to analyse the expression levels of CLCA2 mRNA in normal and tumourous tissue samples, as well as in tumour cell lines and xenografts thereof by utilising Boehringer Ingelheims proprietary BioExpress database (GeneLogic Inc.), which includes expression data from ~ 14.000 human tissue samples, analysed on Affymetrix GeneChips. Although different types of GeneChips have been used for the expression profiling, no statistically relevant variation was observed in the mRNA levels of the corresponding sample set among different GeneChips.

Analyses of the expression pattern in the BioExpress database (GeneLogic Inc.) indicated a preferential overexpression of CLCA2 not only in SCCs of the lung, but also in SCCs of the cervix, the larynx, the head and neck region, in lymph node metastasis of SCCs, in mycosis fungoides a cutaneous T cell lymphoma and in a subset of infiltrating ductal breast cancers.

Interestingly, adenocarcinomas of the esophagus show only minimal levels of CLCA2 transcript in comparison to their corresponding normal esophageal tissue, whereas laryngeal SCC maintain the high expression level of CLCA2 transcript throughout transformation. It cannot be excluded, that the relatively high CLCA2 transcript levels in normal esophageal and laryngeal tissue stems from the presence of an adjacent SCC and/or an inflammatory disease due to the fact that often the "normal" tissue is excised from the adjacent tissue of a tumour. Also a parakrine induction of CLCA2 in the surrounding tissue cannot be excluded. The uncategorised head and neck SCCs show an elevation of CLCA2 transcript levels in comparison to the corresponding normal tissue samples.

Comparison of the expression pattern of CLCA2 in esophagus with that in other tissues of the digestive tract, exhibits no additional expression in other parts of the digestive tract, underlining the restricted expression pattern of CLCA2.

In breast cancer a subpopulation of the infiltrating ductal carcinomas shows overexpression of CLCA2. It remains unclear whether distinct chromosomal aberrations or various differentiation stages lead to the induction of and/or dependence on CLCA2 in this subset of breast cancer samples. In this context it has to be mentioned that Her2 amplification is also found only in a subset of breast cancer samples [150-152].

In the view of elevated CLCA2 transcript levels in tumour-associated skin biopsies, it is of interest that a dramatic increase of CLCA2 mRNA was identified in mycosis fungoides, which is the most common type of cutaneous T cell lymphoma. It is a Non-Hodgkin lymphoma, which is predominantly related to skin. It generally affects the skin, but may progress internally over time. Disease may also progress involving nodes, blood and internal organs, or transform into a higher-grade lymphoma. As shown in this study, expression of CLCA2 seems to be induced in tumour-associated skin rather than in normal skin. Similarly, mycosis fungoides might make use of CLCA2 for its colonisation of the skin during its course of progression. In contrast, no CLCA2 expression is found in Hodgkin lymphomas.

To exclude relevant contribution of cells of the tumour stroma, the CLCA2 expression pattern was also investigated in several laser-capture-microdissected (LCM) tissue samples of lung carcinomas in comparison to non-neoplastic lung

DISCUSSION

tissue samples. Based on data from the *in silico* analysis, an expression of CLCA2 in tumour stroma could be excluded and the tumour-specific expression pattern of CLCA2 was confirmed.

It seems that formation of SCC-derived metastases are somehow linked with CLCA2 expression, supporting the idea that CLCA2 might play a crucial role in the formation of SCC-derived metastases and being essentially involved in the progression of mucosis fungoides. In contrast to mycosis fungoides no (elevated) expression of CLCA2 was detectable in B cell and T cell lymphomas and their corresponding normal tissues/cells.

The analysis of CLCA2 transcript levels in lymph node metastases from different primary tumour subtypes showed, that SCC cells, which metastasise into the lymph node, maintain their high CLCA2 transcript levels during the infiltration of the lymph node. Again a highly specific expression pattern was found as lymph node metastases descending from primary melanomas and adenocarcinomas do not express CLCA2.

All SCC tissues with elevated levels of CLCA2 transcript (as compared to their corresponding normal tissues) do not show overexpression in their corresponding adenocarcinomas; CLCA2 overexpression is clearly restricted to SCCs.

By comparison of the CLCA2 expression level in breast cancer specimens with that in their corresponding metastases, it was found that CLCA2 overexpression in breast cancer seems to be needed for formation of brain-, lung-, liver-, and lymphnode-metastases (see Figure 16). A correlation between CLCA2 overexpression and the development of metastases, was also indicated by analysis of expression data derived from cell lines and their corresponding developing tumours in xenograft models. All xenografts originating from SCC cell lines such as A431 (cervix carcinoma), H520 (lung carcinoma) and HNOE (head and neck/esophagus carcinoma) exhibit a dramatic upregulation of CLCA2 when xenografted into nude mice (Figure 14). Except for the pancreatic tumour cell line BxPC3 and MDA-MB-453 (breast cancer cell line) none of the adenocarcinoma-derived cell lines lead to such a pronounced upregulation of CLCA2 in xenografts. In the case of MDA-MB-453 a high expression level of CLCA2 is already detectable

DISCUSSION

in 2D culture (Figure 14). Altogether, these data support the idea that CLCA2 plays a crucial role in tumour progression.

In addition, correlation of CLCA2 expression with receptor- and lymph node status in human breast cancer samples reveals that CLCA2 expression is mainly found in estrogen receptor-negative or in Her2-positive specimens.

Analysis of a panel of different human carcinoma cell lines again underlines the restricted expression of CLCA2 as it is limited to SCC or human breast cancer cell lines (Figure 15).

Analysis of chromsomal amplifications of the CLCA2 locus on Affymetrix SNP-Chips in human tumour cell lines showed that most SCC cell lines with an increased expression level of CLCA2 also exhibit amplification of 1p22-31 [63]. A copy number of 3 of the CLCA2 loci was found e.g. for FaDu, a head and neck SCC cell line, possibly reflecting the link between genomic and expressional status in various SCCs.

In statistical analyses – correlating normalised expression values of the BioExpress database (GeneLogic Inc.) – the next step was to search for transcripts exhibiting a similar expression pattern. Among those genes various markers of stratified epithelia, such as desmocollin (DSC3), desmoglein 3 (DSG3), TP53 apoptosis effector (PERP), plakophilin 1 (PKP1), and keratin 5 (KRT5) were extracted. The expression of those epithelial differentiation markers seems to be maintained during the transition from a stratified epithelial cell into a hyper-proliferating SCC cell, and could therefore be essential for the establishment of SCCs. Further identified genes, which were already described to be involved in tumourigenesis, are: Ladinin 1 (LAD1/CD18), tripartite motif-containing 29 (TRIM29), stratifin (14-3-3 sigma/SDN), or RAB38, a member of the RAS oncogene family. Another set of genes with a similar expression profile may represent yet unidentified markers for stratified epithelia and/or genes involved in the tumourigenesis of SCCs. One of these genes, which showed a more or less identical expression profile is the G-protein-coupled receptor GPR87. Its association with stratified epithelia and essential contribution to SCC carcinogenesis was shown recently by S. Glatt, a PhD student of our group [79]. Whether GPR87 and CLCA2 are directly interacting

DISCUSSION

with each other is not known yet. However, interaction of GPCRs and ion channels for regulation of signal transduction has already been demonstrated [153;154].

Knock-Down of CLCA2 in CLCA2-Positive Tumour Cells

With loss-of-function studies by siRNA-mediated knock-down of CLCA2, the implication on cell survival and proliferation of CLCA2 was investigated. In proliferation assays could be shown that downregulation of CLCA2 in HN5 cells (SCC cell line) leads to inhibition of proliferation, to reduction in viability and subsequently to induction of apoptosis. The same holds true for GPR87 [79], which exhibits a nearly identical expression profile as CLCA2 and was used as a control target gene for siRNA experiments. Downregulation was monitored by RT-qPCR. Testing different siRNA oligonucleotides, the two most efficient ones (siCLCA2-2 and siCLCA2-4) were selected for further studies. In general, siRNA oligonucleotide siCLCA2-4 lead to the most pronounced biological effects in all experimental settings.

Investigation of various apoptosis parameters by a high content screening approach (Cellomics) utilising siRNA-transfected HN5 cells revealed dramatic changes in the membrane permeability and in the mitochondrial status. In addition, a pronounced nuclear fragmentation could be observed, which supports data obtained from PARP-cleavage experiments on Western blots reflecting induction of apoptotic events.

CLCA2 knock-down in siRNA-treated HN5 cells also lead to a significant shift in the cell cycle. The cell count in G1-phase (2n) was clearly reduced, and cells were arrested in G2, whereby polyploidy increased. Most likely apoptosis was already initiated in these polyploidic cells. Altogether, these data indicate that both, CLCA2 and GPR87 play an essential role for proliferation of HN5 cells.

As MDA-MB-453 cells (breast cancer cell line) exhibit much higher CLCA2 expression levels than HN5 cells, they were also used in siRNA experiments. Interestingly, in contrast to HN5 cells, MDA-MB-453 cells were insensitive towards knock-down of CLCA2 when tested in a proliferation assay.

Furthermore, no PARP-cleavage could be detected in siRNA-treated, CLCA2-constitutively-expressing T47D clone #47. T47D is also a breast cancer cell line which – in contrast to MDA-MB-453 – has only a hundred-fold lower expression level for CLCA2.

In contrast to the SCC cell lines HN5, the breast cancer cell line MDA-MB-453 (although exhibiting high levels of CLCA2 transcript) seems not to be dependent on CLCA2 for proliferation and survival in 2D-culture. Also gain-of-function in the breast cancer cell line T47D did not change the growth behaviour or generates a sensitivity towards siRNA-mediated downregulation of CLCA2. As both breast cancer cell lines are derived from adenocarcinomas, we conclude that downregulation of CLCA2 leads only to severe effects in cell lines of SCC origin (as e.g. HN5).

Antibody Generation and Immunohistochemistry

Due to the lack of commercially available CLCA2-specific antisera it was repeatedly tried to generate specific antisera against human CLCA2. Rabbits were immunised with peptides derived from identified CLCA2-specific epitopes which discriminate between other members of the CLCA family (for details see "Results").

Fractions of affinity-purified sera of immunised rabbits were tested on Western Blot, but none of these antibodies generated a specific signal. Therefore, it was decided to recombinantly introduce an HA-tag to the wild type CLCA2 coding sequence in order to allow detection of CLCA2 expression at least in stably or transiently transfected cells utilising an anti-HA-antibody. The time-dependent induction of inducible HA-tagged CLCA2 expression could be confirmed with HA-specific antibodies in stably transfected T47D.

Interestingly, one of the affinity-purified antisera, which was not suitable for detection of CLCA2 on Western blots but exhibiting a high titer towards peptide K94-N110 of CLCA2, showed a specific staining of tumour cells in human SCC lung tumour samples confirming its tumour-specific expression (tumour stroma and normal tissue is negative). A pre-immune serum was used as a negative control. The signal could also be competed with a peptide which was applied for generating the antiserum. Staining of adenocarcinomas, however, are not conclusive yet, as immunohistochemical analyses of these tumour types lead to aggregation of the antiserum. Therefore, these data have to be regarded as preliminary, also because (due to availability) only a small number of human SCCs have been analysed yet.

Proteolytic Processing

Putative Hydrolase Domain

Pawlowski and colleagues published in 2006 that CLCA1 contains a hydrolase domain which is responsible for the processing of mature full-length CLCA1 [65]. Detailed sequence comparison revealed that the homology within the CLCA family is quite high. Making use of the etstablished inducible expression of HA-tagged CLCA2 in T47-D cells, it was first tried to detect a C-terminal fragment on Western blots. Indeed, the cleavage product could only be detected upon Tet-induced CLCA2 expression in a time- and concentration-dependent manner. Non-induced cells under the same conditions did not show any specific band on Western blots (Figure 41, Figure 43). An alignment of CLCA1 and CLCA2 protein sequence exhibited a high homology within the segment where the proposed hydrolase is located (in CLCA2 at AA 29-208). As the essential residues which might be involved in the formation of the predicted hydrolase domain are conserved, it was started to analyse the proteolytic activity of CLCA2 in more detail. Experiments with matrixmetalloprotease-inhibitors such as Phenantrolin, Diprotin B, Bestatin, Amastatin and others did not allow to draw any valid conclusions as most of these compounds are quite toxic and can only be used at very low and thus uneffective concentrations in tissue culture. Next, all conserved Cysteine residues in the coding region of CLCA2 which are located within the region where the putative hydrolase domain is embedded were exchanged. Mutation of conserved Cysteines at position C211 and C216 still allowed a residual proteolytic activity to take place although at a very low level. Mutations at C132, C196, and C206 completely abolish the proteolytic processing (Figure 46, Figure 48). Whether the inhibitory effects of the respective mutations are due to the influence on the correct formation of the active site or are due to a more general influence on the overall structure and/or stability is not clear yet. C132 corresponds to the Cysteine residue which is suggested to keep H140 (equivalent AA in CLCA1: H133) the proposed structural metal site (Zn^{2+}

coordination) in correct orientation. Next, the corresponding sites in CLCA2 which might be directly involved in the formation of the catalytic site were mutated. The exchange of basic Histidines H164 and H168 with Cysteine (polar), Tyrosine (polar) or Asparagine (polar) completely inhibits cleavage of CLCA2. Exchange of Glutamic acid E165 with Aspartic acid or Glutamine also completely inhibits cleavage. Based on the modelling data from Pawlowski et al. [65] these residues are supposed to take part in the formation of the putative catalytic site. These results strongly supports the presence of a zincin-like hydrolase domain in the N-terminal region of CLCA2 as well. Furthermore, by generating hydrolase dead mutants it could be shown that the proteolytic activity is an autocleavage rather than a trans cleavage event managed by an additional cellular protease. Whether this autocatalytic activity is performed in cis or trans is still open. However, to clearly demonstrate a 1/1 molar ratio of the CLCA2 protein and Zn^{2+} further experiments are needed [155].

Putative Cleavage Site

As known from bioinformatic analyses and literature [95], a mono basic cleavage site is predicted at position AA674. Mutational analysis by "Alanine-walking", starting from predicted positions P5 to P4' (AA670 to AA678) by subsequent and stepwise substitution of the respective amino acid to an Alanine, were performed. Additionally, at those positions where the cleavage activity could be abolished after introduction of an Alanaine, additional exchanges were performed at the respective position. In particular at positions P3, P1, and P1' all classes of amino acids i.e. amino acids with basic, acidic, non-polar or uncharged polar side chain) were introduced, in P2' two classes (non-polar and uncharged polar side chain). These positions have been chosen as they are in most cases the cleavage specificity-determining amino acids [156]. Data from this study support this idea as predicted position P1 (AA674) strictly not allows any exchange of Arginine with others than Lysine (both AA contain a basic side chains). According to the mutational analysis the cleavage recognition site might be as follows:

P3(Tyr/Phe) – P2(X) – P1(Arg/Lys) – P1'(Tyr/Phe) – P2'(Phe/Tyr).

DISCUSSION

Inhibition of Post-Translational Modifications

For investigation of post-translational modifications of CLCA2 (glycosylation, isoprenylation) and their contribution to stabilisation and/or proteolytical processing, expression of CLCA2 was studied in the presence of the glycosylation inhibitor Tunicamycin and the isoprenylation inhibitor Perillic acid. By utilising the T47D Tet-on inducible expression system for wt-CLCA2 it could be shown that both, the stability and the proteolytic processing are impaired when glycosylation is inhibited. How glycosylation influences the post-translational processing is unclear. Two scenarios are conceivable: Either impaired glycosylation leads to incorrect folding and protection of the cleavage site or leads to a misguided cellular localisation where proteolytic processing cannot take place. Another explanation would be that the cleavage product of non-glycosylated CLCA2 becomes degraded and therefore cannot be detected on Western blots.

Based on these data it was decided to mutate the putative glycosylation sites in CLCA2. The use of several glycosylation site prediction software tools resulted in prediction of the putative glycosylation sites N150 (located within the predicted hydrolase domain) and N822. Point mutations resulting in exchange of Asparagine to Alanine of these two sites had no influence on the cleaving efficacy. However, mutating N822 resulted in a shorter C-terminal fragment after cleavage, indicating that an incorrect positioning of the cleavage site (e.g. triggered by loss of gylcosylation) leads to processing at another site with a related amino acid motif.

Integrin β4

Abdel-Ghany found Integrin β4 to be coprecipitated with CLCA2 [70]. He also identified a strong correlation between the level of Integrin β4 and adhesion of HEK293 cells expressing CLCA2, and a correlation between Integrin β4 expression in tumourigenic cell lines *in vivo* and the number of lung metastases detected in nude mice [70]. Upregulated Integrin β4 levels were also found in microarray studies of lung cancer metastases [74].

Based on data of expression profiling on Affymetrix GeneChips, it could also be identified an enhanced Integrin β4 expression upon induction of CLCA2 expression. Whether the upregulation leads also to an enhanced activation status of Integrin β4 or not, could not be answered due to the lack of any functional phosho-specific antibodies for Integrin β4. As members of the integrin protein family are known to be involved in cell attachment to proteins of the extracellular matrix, integrins therefore play an important role in metastasis, regulating not only dissemination, but also homing of metastasising tumour cells [157]. To gain a more detailed understanding about the specific induction of integrins upon induced CLCA2 expression, integrin-mediated cell adhesion assays were performed. The "Integrin-Mediated Cell Adhesion Arrays" (Chemicon) are efficient tools to screen cell surface protein profiles. The method is based on stripwells with each well pre-coated with an antibody detecting an individual integrin or extracellular matrix protein. Utilising the established inducible expression system for CLCA2, it was possible to screen for induction of α- and β-integrins in dependence on induction of wt- or mutant-CLCA2 expression in T47D Tet-on cells (control: T47D cells which stably carry the "empty" expression vector). Out of the α-integrin family no specific induction could be detected, whereas for β-integrin family member B2 a specific induction upon CLCA2-induced expression could be demonstrated (Figure 56, Figure 57).

3D-In Vitro Carcinoma Assay: Spheroids

For a better simulation of the situation in a tumour, CLCA2-expressing clones were also grown in multicellular tumour spheroids, which are *in vitro* 3D models that recapitulate malignant cell contacts within a tumour. Multicellular tumour spheroids can for instance also be used to evaluate the tumour's response to therapeutic agents or the interaction with cells of the tumour stroma, e.g. the von Willebrand factor domain in CLCA2 was previously shown to have a role in adhesion by binding at Integrin β4 [158] and preliminary experiments confirmed at least the interaction with members of the β-integrin family (see above).

In the corse of this study the *in vitro* 3D model [159] was used for analysis of the influence of expression of active and inactive CLCA2 on three-dimensional cell growth and viability. Since not all cell lines or cell types are capable of forming spheroids, first a cell line capable of spheroid formation had to be selected. Fortunately, cell line T47D which was used to establish constitutive and inducible

CLCA2-expressing clones could be cultivated as spheroids. Stable spheroids were transferred to a collagen gel or Matrigel and grown until they became necrotic and/or apoptotic due to the lack of supply with nutrients and oxygen in the centre of the 3D spheroid, which are normally provided by *de novo* formed vessels, docking the growing tumour to the existing blood vessel system of the organism (neoangiogenesis).

In a statistically meaningful setting the relative cell growth of 2D- (counting viable cells) with that of 3D-cultures (determination of area [µm2]; Figure 63) was compared over a period of 3 to 10 days. Although, the relative cell growth of CLCA2-expressing T47D Tet-on clone #3 in induced versus non-induced status did not show a difference in 2D-cultures, it indeed exhibited a pronounced difference (> 2.5-fold) between 2D- and 3D-culturing. In contrast to CLCA2-inducible clone #3, the constitutive clone #47 showed already an enhanced relative cell growth in 2D-culture which was similar to that found in 3D-cultures. In general, constitutive CLCA2 expression during spheroid-formation lead to generation of larger spheroids and subsequently to enhanced apoptosis due to a higher proliferation rate. Without using the 3D-spheroid model, detection of the essential contribution of CLCA2 for a three-dimensional growth wouldn't have been possible. It is also of interest that CLCA2 mutants defective in their proteolytic processing apparently had no influence on spheroid formation or growth properties of spheroids. Therefore, maturation processing of CLCA2 seems not to have an impact on the three-dimensional growth of this cell line.

Affymetrix Microarray Studies

To study the effects of overexpression of CLCA2 on gene expression, expression profiling using the Affymetrix GeneChip Human Genome U133 Array was performed by comparing cell clones with induced expression of CLCA2 with their uninduced counterparts. In a separate experimental setting, the expression pattern for selected transcripts was confirmed by RT-qPCR. Among those (and besides CLCA2, which served as an internal control), genes such as CDIPT, MLL2 and SFI1 were identified. CDIPT (CDP-diacylglycerol-inositol 3-phosphatidyltransferase; phosphatidy inositol synthase) is a gene whose inhibition causes small cell lung

DISCUSSION

carcinoma cells to arrest in G1 [98]. It further has been suggested to be involved in oral carcinogenesis [99]. MLL2 (myeloid/lymphoid or mixed-lineage leukemia 2) plays a role in apoptosis and in alterations of cell adhesion. MLL2 is a member of the human MLL family, which belongs to a larger SET1 family of histone methyltransferases. MLL2 is a transcriptional activator that induces the transcription of target genes by covalent histone modification. MLL2 appears to be involved in the regulation of adhesion-related cytoskeletal events, which might affect cell growth and survival [100]. Members of the MLL protein family have been found to cause acute leukemia. Recently, deletion studies revealed that alterations of the CxxC domain (Zn-binding domain) disrupts its oncogenic potential [101]. SFI1 (SFI1 homolog, spindle assembly associated; yeast) is involved in initiation of centrosome duplication. SFI1 has been identified as a binding partner for the Ca^{2+}-binding protein centrin, suggesting a role in the dynamic behavior of centrosomes. SFI1 binds to multiple centrin molecules and the complex forms Ca^{2+}-sensitive contractile fibers that function to reorient centrioles and alter centrosome structure [96]. So far no correlation to tumour cell growth has been reported. However, according to the *in silico* analysis of this study, SFI1 is found upregulated in many tumours, with highest expression levels found in various B cell lymphomas and B cell leukemias (data not shown).

In total, approximately 150 genes were shown to be differentially up- or downregulated in a statistically relevant mode (fold change > 1.8; p-value < 0,0001). In order to link these expression profiles of regulated genes to canonical pathways and bring them into a biological context, these data were analysed with the Ingenuity Pathways Analysis Software (IPA; of Ingenuity Systems Inc.). This program is a text- and data-mining program which not only links gene and protein functions/interactions based on the knowledge of published data in the most relevant journals, but also allows to bring these interactions into a context of biological functions, regulatory cascades and functional groups (e.g. oncology/metastasis/prostate). It has to be mentioned that the presented studies can only be regarded as preliminary as the outcome of these *in silico* analyses has to be proven in experimental settings. On the other hand these data allow a much more focused proceeding for future experiments as this text- and data-mining program not only describes protein-protein interactions, but also other types of reciprocal action such as protein-nucleic acid interactions.

Based on these *in silico* analyses several interaction-networks were identified.

Network I (Figure 85) indicates that a group of specifically induced genes upon induction of CLCA2 is strongly linked to the interaction-axis of SMAD3–TP53–MYC– HNF4A–ERBB2. It has been shown recently that HNF4alpha1 (HNFA4) and the proliferation factor c-Myc may compete for control of genes involved in cell proliferation and differentiation [160].

Network II (Figure 86) comprises a link to the TNF–IL2–IL10 interaction network. Interestingly, treatment of SCCs of the skin with a Toll-like receptor agonist such as imiquimod contained a decreased percentage of regulatory T (T reg) cells producing less IL10 and TGF-beta, thereby inhibiting their suppressive activity. This means that SCCs evade the immune response at least in part by recruiting T reg cells via modulation of the IL10 pathway [119].

Network III (Figure 87) brings some of the differentially upregulated genes into the context of the ERK–MAPK signaling cascade with ERK as the center of this network. The most significant networks were then merged (Figure 88), for the sake of simplicity focusing exclusively on protein-protein interactions, highlighting biologically interesting clusters such as the cluster of v-src (involved in regulation of many different signaling cascades), HNFA4, ERBB2, and KPNA2. Karyopherin 2 (KPNA2), for instance, is a member of the karyopherin (importin) family, which is part of the nuclear transport protein complex. KPNA2 is responsible for transfer of the repair complex RAD50–MRE11–NBS1 into the nucleus. The cytoplsamic portion of NBS1 promotes tumourigenesis through the binding and activation of the PI3-kinase/AKT pathway which is dependent on KPNA2 [161]. Previously KPNA2 overexpression was identified as a prognostic marker in breast cancer correlating with poor survival [162]. KPNA2 was also shown to be highly upregulated in esophageal cancer [163].

In a next step, the involvement of canonical signaling transduction pathways were investigated by phospho-proteomic studies. To learn more about the contribution of some important executer proteins of canonical pathways, which might be found activated/deactivated (=phosphorylated/dephosphorylated) in experimental settings

DISCUSSION

of CLCA2 induction, cell lysates of induced and non-induced T47D Tet-on clone #3 cells (taken at various time points) were analysed an Western blots utilising phospho-specific antisera.

An increase in phosphorylation of AKT (S473), ERK 1&2 (T185/Y187), MAPK (T202/Y204), RAF (S259), CRAF (S338), and PDK1 (S241) was observed. No change in the phosphorylation pattern was seen for AKT at position S308. These data confirm the predictions made through the usage of the IPA software, indicating a link to the ERK–MAPK canonical pathway. From literature it is known that in oropharyngeal squamous cell cancer patients phopshorylation of AKT at S473 is linked with a poor clinical outcome [164]. The malignant potential of SCC cell lines was shown to be significantly higher in CRAF-expressing cells and CRAF expression maintained also the radiation-resistant phenotype indicating that CRAF plays an important role in the squamous cell carcinogenesis [165]. Phosphoinositide-dependent kinase 1 (PDK1), a downstream target of PI3-K, has been recently suggested to take part in the signaling cascade of gastrin releasing peptide [(GRP)-Src-PI3K-PDK1-TACE-amphiregulin-EGFR] with multiple points of interaction leading to induction of proliferation of human head and neck squamous cells [166].

Analysed proteins which were found to be dephosphorylated upon CLCA2-induction are: mTOR (p2448), 4E-BP1 (T37/T47), STAT1 (Y701), and NPM (T234/T237). Deregulation of mTOR, 4E-BP1, STAT1 indicates a strong impact of CLCA2 to translational modulation whereas nucleophosmin (NPM) is involved in cytoskeletal regulation [120].

The phosphorylation and dephosphorylation upon CLCA2-induction of selected proteins was also analysed with the "Human Phospho-MAPK Array Kit" and "Human Phospho-RTK Array Kit" of R&D Systems. Performing these assays, a higher phosphorylation and therefore activation level for ERK1 and ERK3 could be confirmed in CLCA2-constitutively-expressing T47D clone #47 compared to control cells (empty vector). In addition, a higher phosphorylation level of EGF-R, Insulin-R, and IGF-I R was found in T47D clone #3 cells upon induction of CLCA2. The biological meaning of these findings is supported by recent publications e.g. showing that a worse tumour differentiation and a positive nodal stage in head and neck SCC are significantly associated with EGFR overexpression. In contrast to NSCLC, protein overexpression rather than mutation might be responsible for activation of the EGFR pathway in head and neck SCC [167].

Summary

In the context of targeted therapy, ion channels are representing promising players. However, very little is known about their specific contribution to tumourigenesis. In the course of this Thesis a preferential overexpression of CLCA2 in human squamous cell carcinomas (SCC) of different origins including their lymph node metastases could be demonstrated. In addition, overexpression was also detected in mycosis fungoides, a cutaneous T cell lymphoma and in a subset of infiltrating ductal breast cancers. Preliminary IHC analyses confirmed expression in human SCC samples. *In silico* expression analyses identified a dramatic increase in CLCA2 expression when SCC cell lines are xenografted into nude mice supporting that CLCA2 plays an important role in late stage tumourigenesis.

Loss-of-function studies in SCC cell lines by siRNA-mediated knock-down of CLCA2 leads to inhibition of proliferation, to reduction in viability, subsequently to induction of apoptotic processes, and to an arrest in G2. In contrast, breast cancer cell lines stayed unaffected at least in 2D-tissue culture. Inducible expression of CLCA2 in tumour cells in combination with a detailed mutational analysis demonstrated a proteolytic (most likely autocatalytic) processing (hydrolase) activity of CLCA2 and proposed a structural metal site for Zn^{2+} coordination. Altogether, these data strongly supports the presence of a zincin-like hydrolase domain in the N-terminal region of CLCA2. As predicted by bioinformatic analyses and in analogy to CLCA1 a mono basic cleavage site was predicted and could be confirmed by "Alanine-walking" mutagenesis with a predicted cleavage recognition site of P3(Tyr/Phe) – P2(X) – P1(Arg/Lys) – P1'(Tyr/Phe) – P2'(Phe/Ala). Inhibition of glycosylation revealed that both, the stability and the proteolytic processing are impaired when glycosylation is inhibited.

For a better simulation of the situation in a tumour, CLCA2-expressing clones were also grown in multicellular tumour spheroids. In statistically meaningful experimental settings, the relative cell growth rate of 2D- and 3D-cultures was compared and indeed a > 2.5-fold increase in proliferation in the three-dimensional system could be demonstrated upon induction of CLCA2 expression.

For a better understanding of the involvement of CLCA2 in signal transduction a comprehensive expression profiling study on Affymetrix GeneChips was performed, followed by RT-qPCR and phospho-proteom studies with selected genes. In total ~ 150 genes were shown to be differentially regulated upon induction of CLCA2 expression. Among those genes are CDIPT, MLL2, and SFI1, which play an important role in cell cycle regulation. Utilising Ingenuity Pathways Analysis – a text- and data-mining program – several interaction-networks were identified. All of them could be linked with important canonical signal transduction cascades or interaction networks such as the TNF–IL2–IL10 interaction network or the ERK–MAPK signaling cascade. Phospho-proteom studies confirmed these findings and additionally identified a strong impact of CLCA2 on translational modulation.

Although these expression profiling data can only be regarded as preliminary, they however, offer the basis for further focused studies for the characterisation of CLCA2 as a promising target in SCC tumourigenesis.

MATERIAL AND METHODS

MATERIAL AND METHODS

Molecular Biology

Total RNA Isolation

The medium was removed and cells were directly lysed on the plates using 1 ml Trizol reagent (Invitrogen, Cat.# 15596-018) per 10 cm² of tissue plate. To ensure total lysis of the cells the suspension was pipetted up and down several times. The lysate was stored at -20°C, or used directly for RNA isolation.

For RNA isolation, the lysate was incubated for at least 5 min at RT (room temperature) to ensure complete dissociation of nucleoprotein complexes. 0.2 ml Chloroform ($CHCl_3$) per ml Trizol was added, vortexed for about 15 sec and incubated for 2-3 min at RT. Afterwards the lysate was centrifuged at 10,000 rpm for 5 min at 4°C and the upper aqueous phase was transferred into a fresh RNAse free tube.

The RNA was precipitated using 0.5 ml isopropanol per ml Trizol by incubating at least 10-15 min at 4°C. Subsequently the RNA was spun down by centrifugation at 15,000 rpm for 5 min at 4°C. The RNA pellet was washed with 75 % RNAse-free Ethanol and centrifuged at 15,000 rpm for 5 min at 4°C. The supernatant was then removed and the pellet was air dried.

Finally, the RNA was resuspended in 10 µl DEPC-H_2O and frozen at -20°C or used immediately for single strand cDNA (ss cDNA) synthesis.

Frozen tumour samples were sectioned and subsequently lysed in 200 µl RLT buffer of RNAeasy Kit (Qiagen, Cat.# 74104) and total RNA was isolated following instructions of the provider.

Single Strand cDNA Synthesis (First Strand Synthesis)

For ss cDNA synthesis the extracted total RNA was reverse transcribed using the "Sprint^{TM} PowerScript^{TM} PrePrimed SingleShots with Random Hexamer Primers" from Clontech (Cat.# 639562).

To eliminate putative contaminations with genomic DNA, 1 ug total RNA was digested with 1 U DNAseI for 30 min at 37°C. The reaction was stopped by the addition of 2.5 mM EDTA and incubation at 65°C for 10 min. For ss cDNA synthesis 10 µl DEPC-H$_2$O (di-ethyl-propyl carbonate-treated H$_2$O) was added to the lyophilised kit components and the DNA-free RNA was also added. The reaction was incubated for 60 min on 42°C. The ss cDNA was stored at -20°C, or used for further investigations.

Real-Time Quantitative PCR (RT-qPCR)

For RT-qPCR cDNA samples were prepared following the instructions as described above. Analyses were performed with commercially available gene-specific "TaqMan Gene Expression Assays" of Applied Biosystems for the genes listed in Table 5.

Gene of Interest	Cat.# and Label
CLCA2	Hs00197957, FAM
SATB1	Hs00161515, FAM
SFI1	Hs00826823, FAM
MLL2	Hs00231606, FAM
EP300	Hs00914223, FAM
CDIPT	Hs00197004, FAM
TP53BP1	Hs00996818, FAM
LTBP1	Hs00386448, FAM
TSC1	Hs00184423, FAM
MACF1	Hs00201468, FAM
DTX2	Hs00539707, FAM
CHD6	Hs00260089, FAM
AFF4	Hs00232683, FAM
UNC84A	Hs00391956, FAM
RUTBC3	Hs00210437, FAM
P4HA2	Hs00188349, FAM
C10orf118	Hs00215984, FAM
ASPM	Hs00411505, FAM

Table 5. Genes which were tested in "TaqMan Gene Expression Assays" (Applied Biosystems)

In all gene expression assays, a FAM-dye labelled minor-groove-binder probe linked to a non-fluorescent quencher is used in combination with two non-labelled flanking primers. Probe and primers are sequence-specific for the respective gene of interest. The fluorescent signal is generated due to the release of the quencher from the FAM-labelled probe after the induction of DNA-synthesis from the two flanking primers.

As endogenous control the "Taq Man Gene Expression Assays" of the "housekeeping gene" β-2-microglobulin (Applied Biosystems, Cat.# 4326319E) was used, which is VIC-labelled and can be measured parallel to the FAM-channel.

For all assays the "Taq Man Gene Expression Master Mix" (Applied Biosystems, Cat.# 4369016) was used. Quantification and analyses were performed by the Mx4000 device from Stratagene, using the provided software package.

For RT-qPCR the following reagents were combined:

reagent	µl per reaction tube
cDNA-sample	2
2x TaqMan Gene Expression Master Mix	12.5
20x target primers and probe	1.25
20x control primers and probe (of B2M)	1.25
nuclease-free water	8
total volume	25

Cycling Parameters for RT-qPCR using MX4000, Stratagene:

1 cycle:	10 min	95°C
40 cycles:	15 sec	95°C
	1 min	60°C
hold at:		4°C

PCR (Polymerase Chain Reaction)

PCR-Amplification

This method was used to amplify specific DNA-fragments by using rapid changes in temperature and a DNA-Polymerase in connection with dNTPs and specific primers for the DNA sequences of interest.

For amplification of CLCA2, the commercially available vector „Human Full-Length cDNA Clone in pCMV6-XL4" from Origene (Cat.# TC116023, Acc. NM_006536: Human calcium-activated chloride channel, family member 2 (CLCA2)) was used as template.

For the amplification reaction the following reagents were combined:

reagent	µl per reaction tube
aqua dest.	30
10x Pfx amplification buffer	5
10 mM dNTP mixture	1.5
PCR primer forward (5 pmol/µl)	5
PCR primer reverse (5 pmol/µl)	5
50 mM $MgSO_4$	1
DNA template (plasmid: 10 ng/µl)	2
Platinum Pfx DNA polymerase (2.5 U/µl)	0.5
total volume	50

The following amplification parameters were designed to be used in a "PE GeneAmp PCR System 2400" or "PE GeneAmp PCR System 9700":

1 cycle:	2 min	94°C
30 cycles:	15 sec	94°C
	30 sec	55°C
	3 min	68°C
1 cycle:	10 min	68°C
hold at:		4°C

Aliquots of 5 µl per reaction were analysed by electrophoresis on ethidiumbromide-(EtBr) stained 4 % agarose gels and amplified DNA-fragments were visualised on an UV-transilluminator.

Colony Screening

Colony PCR screening was used for rapid determination of positive transformants. Bacterial colonies were picked from the agar plate, boiled for 10 min in 100 µl TE, either in water bath or in a thermocycler. The suspension was cooled down on ice and 2 µl suspension was pipetted into a PCR tube for amplification.

The following reagents were combined:

reagent	µl per reaction tube
aqua dest.	12.2
10x PCR buffer	2
dNTP mix (2.5 mM each)	1.6
PCR primer down (5 pmol/µl)	1
PCR primer up (5 pmol/µl)	1
boiled E. coli solution	2
Taq polymerase (5 U/µl)	0.2
total volume	20

The following amplification parameters were chosen in a "PE GeneAmp PCR System 2400" or "PE GeneAmp PCR System 9700" with a minimum of 1 min of elongation time and per additional 1000 bp one extra min for elongation.

1 cycle:	3 min	94°C
28 cycles:	1 min	94°C
	1 min	55°C
	1 min	72°C
1 cycle:	7 min	72°C
hold at:		4°C

Aliquots of 5 µl per reaction were analysed by electrophoresis on EtBr-stained 4 % agarose gels and amplified DNA-fragments were visualised on an UV-transilluminator. Colonies corresponding to samples, which showed an amplification product of the correct size, were picked and grown in 2 ml respective medium o/n and used for plasmid purification.

Cloning PCR Products (TA-Cloning)

The method is based on the non-template-dependent activity of Taq polymerase which adds a single desoxadenosine to the 3'-end of a PCR product. The fragment can then be ligated to a vector (pGEM-T Easy Vector of the "pGEM-T Easy Vector System" (Promega, Cat.# A1380)) containing a desoxythymidine residue overhang. Usually, 3 µl of fresh PCR product (10-20 ng/µl) was directly ligated to the pGEM-T Easy Vector (50 ng/µl).

If the PCR product was amplified via a polymerase with proof-reading activity (e.g. Platinum Pfx from Invitrogen), a desoxadenosine had to be added to the 3'-end of the purified PCR product (see "A-Tailing of PCR-Products").

The ligation reaction was performed o/n at 14°C into 50 ng vector in a total volume of 10 µl containing 1 µl 10x ligation buffer, 1 µl pGEM-T Easy Vector (50 ng) and 1 µl T4 DNA ligase (3 U/µl). For highest efficacy the molar ratio (vector : PCR product) should be between 1:1 and 1:3. The ligation product was immediately used for transformation or stored at -20°C.

A-Tailing of PCR Products

To add a desoxadenosine to the 3'-end of the purified PCR product, which is relevant if such PCR product should be directly cloned to the pGEM-T Easy Vector, the following reagents were combined:

reagent	µl per reaction tube
purified DNA-sample	7
dATP 2 mM	1
10x buffer	1
rTaq-polymerase (e.g. TakaRa) 5 U/µl	1
total volume	10

The reaction mixture was incubated for 30 min at 70°C and then directly ligated to the pGEM-T Easy Vector. For this, the following reagents were combined:

reagent	µl per reaction tube
A-getailte DNA-Lsg.	7
pGEM-T Easy Vector	1
10x Ligation buffer (Fermentas)	1
T4-Ligase (Fermentas)	1
total volume	10

The ligation was performed at 14°C o/n.

Restriction Digestion

DNA was digested with restriction endonucleases for about 2 h at 37°C in the appropriate buffer with 3-fold excess (3 U/µg DNA) of restriction enzyme; standard enzymes from e.g. Fermentas, New England Biolabs, Invitrogen were used.

After digestion the enzymes were heat-inactivated at 70°C for 10 min.

The reaction mixture was directly used for further reactions, purified with "QIAquick PCR Purification Kit", or the DNA was precipitated or separated on agarose gels.

Filling 5'-Overhangs with dNTPs

The 5'-3' polymerase activity of "DNA Polymerase I Large (Klenow) Fragment" can be used to fill in 5'-protruding ends, which result from restriction digestion with dNTPs. The optimal reaction conditions are:

1-4 µg	digested DNA
50 mM	Tris-HCl (pH 7.2)
10 mM	$MgSO_4$
0.1 mM	DTT
40 µM	of each dNTP
20 µg/µl	acetylated BSA
1 U/µg DNA	Klenow fragment

Klenow fragment is active in many restriction enzyme buffers, so in most cases the heat-inactivated reaction mix from restriction digestion was directly taken for filling, 1.5 µl dNTP mix (2.5 mM each) and 1 µl Klenow fragment (5 U/µl) were added.

In some cases DNA was first purified by phenol/chloroform extraction and ethanol precipitation or by using the QIAquick PCR purification Kit. After resuspending DNA the Klenow 10 x Buffer, 1.5 µl dNTP mix (2.5 mM each) and 1 µl Klenow fragment (5 U/µl) were added.

The reaction mix was incubated at RT for 10 min.

The reaction was stopped by heating the mixture for 10 min at 75°C.

Converting a 3'-Overhang to a Blunt End

T4 DNA Polymerase exhibits 5'-3' polymerase activity, and 3'-5' exonuclease activity as well, and can be used for the generation of blunt ends from DNA molecules with 3'-overhangs, or for fill 5'-protruding ends with dNTPs.

For converting a 3'-overhang to a blunt end, DNA was digested in a volume of 50 µl with a restriction enzyme that generates a 3'-overhang.

The optimal reaction conditions are:

0.5-2 µg	digested DNA
33 mM	Tris-Ac (pH 7.9)
66 mM	KAc
10 mM	$MgAc_2$
0.5 mM	DTT
0.1 mg/ml	BSA
5 U/µg DNA	T4 DNA polymerase

T4 DNA Polymerase can also be used directly in restriction enzyme buffer.

In most cases 2 µl T4 DNA Polymerase (3 U/µl) was added directly to the heat-inactivated reaction mix of the restriction enzyme digestion.

This reaction mix was then incubated at 37°C for 5 min.

The reaction was stopped by heating the mixture for 10 min at 75°C.

In Vitro Site-Directed Mutagenesis

Site-specific mutations in double stranded (expression-) plasmids were inserted by using the „QuikChange II XL Site-Directed Mutagenesis Kit" from Stratagene (Cat.# 200522).

The method is based on the utilisation of *PfuUltra* high-fidelity (HF) DNA polymerase for mutagenic primer-directed replication of both plasmid strands. The basic procedure utilises a supercoiled double-stranded DNA vector with an insert of interest and two synthetic oligonucleotide primers, both containing the desired mutation. The primers, each complementary to opposite strands of the vector, are extended during temperature cycling by DNA polymerase, without primer displacement. Extension of the oligonucleotide primers generates a mutated plasmid containing staggered nicks. Following temperature cycling, the product is treated with *DpnI*, an endonuclease specific for methylated and hemimethylated DNA, which is used to digest the parental DNA template and to select for mutation-containing synthesised DNA. The nicked vector DNA containing the desired mutations is then transformed into competent cells.

The primer length should be between 25 and 45 bases, with a melting temperature (Tm) of ≥ 78°C. The desired mutation should be in the middle of the primer with ~ 10-15 bases of correct sequence on both sides. The primers should have a minimum GC content of 40 % and should terminate in one or more C or G bases.

For mutant strand synthesis reaction the following reagents were combined:

reagent	µl per reaction tube
10x reaction buffer	5
template (50 ng/µl)	1
forward Primer (5 pmol/µl)	2.5
reverse Primer (5 pmol/µl)	2.5
dNTPmix	1
QuikSolution reagent	3
H$_2$O	34
PfuUltra polymerase (2.5 U/µl)	1
total volume	50

The following amplification parameters were chosen in a "PE GeneAmp PCR System 2400" or "PE GeneAmp PCR System 9700":

1 cycle:	1 min	95°C
18 cycles:	50 sec	95°C
	50 sec	60°C
	9 min	68°C
1 cycle:	7 min	68°C
hold at:		4°C

1 µl of the *Dpn*I restriction enzyme (10 U/µl) was added directly to each amplification reaction and incubated for 60 min at 37°C. 2 µl of the *Dpn*I-treated amplified DNA was transformed to *E. coli* (for details see "Transformation of Bacteria"). The DNA of at least 3 colonies was isolated by "Miniprep" and sequenced.

Purification of DNA

For purification of DNA, the "QIAquick PCR Purification Kit" of Qiagen (Cat.# 28104) was used. This method is adapted to purify single- or double-stranded DNA-fragments from PCR and other enzymatic reactions. Fragments ranging from 100 bp to 10 kb are purified from primers, nucleotides, polymerases, and salts.

5 volumes of buffer PB was added to 1 volume of the sample and mixed. The sample was then applied to a QIAquick column and centrifuged for 30-60 sec at 13,000 rpm. The flow-through was discarded and the column placed back to the same tube. To wash bound DNA, 0.75 ml buffer PE was added to the QIAquick column and centrifuged for 30-60 sec. The flow-through was discarded, the column placed back to the same tube. Then the column was centrifuged for additional 1 min, the flow-through was again discarded and the column placed in a clean 1.5 ml tube. To elute DNA, first 15 µl buffer EB was added to the centre of the QIAquick membrane and after 1 min centrifuged for 1 min. To increase DNA concentration another 10 µl buffer EB was added and after 1 min centrifuged for 1 min. The average eluate volume is 22 µl from 25 µl elution buffer volume.

Phenol/Chloroform Extraction of DNA

This method removes proteins from nucleic acid solutions.

Equal volumes of phenol/chloroform/isoamylalcohol 25:24:1 (v/v/v) and DNA solution were mixed by vortexing for 20 sec and centrifuged 5 min at RT with 15,000 rpm. The upper aqueous phase was transferred to a fresh tube and extracted with 24:1 (v/v) chloroform/isoamylalcohol. The sample was again mixed by vortexing for 20 sec, centrifuged for 5 min at RT and the aqueous upper phase transferred to another fresh tube. DNA was precipitated with ethanol.

Precipitation of DNA/RNA

DNA was precipitated by adding 0.1 volumes of 3 M NaAc, pH 5.2 and 2.5 volumes of ice-cold 96 % ethanol for at least 3 h at -80°C. The most quantitative precipitation will be achieved when incubating at -20°C o/n and centrifugation for 30 min at 15,000 rpm at 4°C. The supernatant was removed and the pellet was washed with 500 µl 70 % ethanol (-20°C) and air dried or briefly dried under vacuum in a vacuum concentrator (Speedvac). The DNA pellet was dissolved in an appropriate amount of elution buffer EB (10 mM Tris-HCl, pH 8.5). For precipitation of RNA the same protocol is valid, but DEPC-treated solutions have to be used.

Alternatively, DNA or RNA can be precipitated with 1 volume of isopropanol without sodium acetate. This is of advantage for large volumes.

Agarose Gel Electrophoresis

Depending on the expected DNA-fragment length agarose-gels ranging from 0.7-4 % agarose were prepared. The agarose was suspended in 1x TBE and shortly boiled in the microwave until the agarose was completely dissolved. The agarose solution was cooled to approximately 50°C and supplemented with EtBr from a stock solution (10 mg/ml) to a final concentration of 1 µg/ml. The mixture was pored into a gel tray and an appropriate comb was stacked into the tray. When the gel was completely solidified the comb was removed and the gel tray was put into the gel chamber containing 1x TBE as running buffer. The samples were mixed in a ratio of 6:1 with 6x loading buffer. Gels were run at 50-200 V depending on the gel size (3-5 V/cm in horizontal configuration).

Recovery of DNA from Agarose Gels

With the "QIAEX II Agarose Gel Extraction Kit" of Qiagen (Cat.# 20021), DNA of 40 bp to 50 kb can be extracted and purified from standard or low-melting agarose gels in TAE or TBE buffers. Up to 400 mg Agarose can be processed per spin column.

The DNA fragments were separated on an agarose gel and fragments of interest were excised with a clean, sharp scalpel. The agarose pieces were transferred into 2 ml Eppendorf tubes and weighted before filling and afterwards. 3 volumes of buffer QG was added to 1 volume of agarose gel and incubated for 10 min in an Eppendorf thermoblock at 50°C to achieve total dissolution of the gel matrix. Then 1 gel volume isopropanol was added and the solution was mixed. A QIAquick spin column was placed in a 2 ml collection tube. The samples were applied onto these columns and centrifuged for 1 min at 15,000 rpm. The flow-through was discarded. To remove all traces of agarose 0.5 ml buffer QG was added and the column centrifuged again for 1 min at 15,000 rpm. The column was washed with 0.75 ml buffer PE. After 2-5 min the column was centrifuged for 1 min at 15,000 rpm. The flow-through was discarded and the QIAquick column was centrifuged for an additional min to remove residual ethanol of the PE buffer. Then the column was placed into a fresh 1.5 ml Eppendorf tube. To elute DNA 30 µl buffer EB was transferred directly on the center of the column. After 1 min incubation, the column was centrifuged for 1 min at 15,000 rpm.

Alternatively to the described method, the "QIAquick Gel Extraction Kit" of Qiagen (Cat.# 28704) was used and performed according to the instructions in the manual.

Ligation of DNA

An insert of interest could be amplified by PCR or "restricted out" from a plasmid containing the insert. Prior to ligation DNA had to be purified using a "QIAquick PCR Purification Kit" or "QIAEX II Agarose Gel Extraction Kit".

150 ng insert was ligated o/n at 14°C into 50 ng vector in a total volume of 10 µl containing 1 µl 10x ligation buffer, 1 µl vector (50 ng) and 1 µl T4 DNA ligase

(3 U/µl). For highest efficacy the molar ratio between vector and insert should be between 1:1 and 1:3.

The ligation product was immediately used for transformation or stored at -20°C.

Transformation of Bacteria

Bacterial Strains

- JM109 High Efficiency Competent Cells with the genotype: *rec*A1, *end*A1, *gry*A96, *thi*, *hsd*R17 (r_K-m_{K+}), *rel*A1, *sup*E44, Δ(*lac-pro*AB), [F´,*tra*D36, *pro*AB, *lacI*^{q}ZΔM15]

  > 10^8 cfu/µg (Promega, Cat.# L2001)

- JM110 Competent Cells with the genotype: *rpsL* (Str^r) *thr leu thi-I lacY galK galT ara tonA tsx dam dcm supE44* Δ*(lac-proAB)* [F´ *traD36 proAB lacI*^{q}*ZΔM15*]

  > 5×10^6 cfu/µg (Strategene, Cat.# 200239)

Transformation of JM109

E. coli cells were thawed on ice.

2 µl ligation reaction was added to an aliquote of 50 µl competent cells and incubated 20 min on ice. Following a 45-50 sec heat shock at 42°C, cells were incubated for further 2 min on ice. Then 950 µl SOC medium (Invitrogen, Cat.# 15544-034) was added and the cells were incubated 1 h at 37°C with shaking (220 rpm). 100 µl of each transformation-culture were plated on agar plates containing the appropriate antibiotic for selection. The plates were incubated o/n at 37°C.

Positive clones were analysed by "Colony Screening (PCR)". The DNA of transformed cells was isolated by "Miniprep" and sequenced.

Transformation of JM110

JM110 *E. coli* cells were used for obtaining plasmids which further needed to be cut with dam- or dcm-sensitive restriction enzymes.

E. coli cells were thawed on ice.

An aliquote of 100 µl of competent cells was mixed with 1.7 µl of β-mercaptoethanol, provided with the kit, and incubated for 10 min on ice, swirling gently every 2 min.

2 µl ligation reaction was added to competent cells and incubated 30 min on ice. Following a 45-50 sec heat shock at 42°C, cells were incubated for further 2 min on ice. Then 950 µl SOC medium (Invitrogen, Cat.# 15544-034) was added and the cells were incubated 1 h at 37°C with shaking (220 rpm). 100 µl transformation-culture was plated on agar plates containing the appropriate antibiotic for selection. The plates were incubated o/n at 37°C.

Positive clones were analysed by "Colony Screening (PCR)". The DNA of transformed cells was isolated by "Miniprep" and sequenced.

Plasmid Preparations

Miniprep

For plasmid preparation, the "QIAprep Spin Miniprep Kit" of Qiagen (Cat.# 27106) was used. The protocol is designed for purification of up to 20 µg of high-copy plasmid DNA from 1-5 ml overnight cultures of E. coli in LB-medium.

E. coli cells were incubated o/n in 2.5 ml LB-amp medium at 37°C in an air-incubator at 225 rpm. The cell suspension was centrifuged for 30 sec at 6,000 rpm. Pelleted bacterial cells were resuspended in 250 µl buffer P1. 250 µl lysis buffer P2 was added and the tube was gently inverted 4-6 times. Then 350 µl buffer N3 was added and the tube was again gently inverted 4-6 times. After 10 min centrifugation at 15,000 rpm, the supernatant was applied to the QIAprep column and centrifuged for 1 min at 15,000 rpm. The flow-through was discarded and the plasmid DNA which was bound to the column matrix was washed once with 0.5 ml buffer PB and once with 0.75 ml buffer PE. Between washing steps, the columns were centrifuged and the flow-through was discarded. After washing steps the column was again centrifuged for 1 min to remove residual wash buffer. Then the QIAprep column was placed into a clean 1.5 ml tube and the DNA was eluted with 50 µl buffer EB (10 mM Tris-HCl, pH 8.5). After 1 min incubation, the column was centrifuged for 1 min.

The DNA-solutions were stored at -20°C.

Maxiprep

For preparation of larger amounts of DNA, the "EndoFree Plasmid Maxi Kit" from Qiagen (Cat.# 12362) was used. The protocol is designed for purification of 500 µg of endotoxin-free plasmid DNA.

300 ml overnight bacteria culture was centrifuged for 30 min at 4°C at 6,000 rpm in a "Sorvall RC-5B Refrigerated Superspeed Centrifuge".

The pellets were resuspended in 10 ml P1 buffer. Then 10 ml P2 buffer was added and the tube was gently mixed. After 5 min incubation at RT, 10 ml of chilled P3 buffer was added and the reaction mixture again gently mixed.

The lysate (but not the plunger) was transferred to the barrel of the QIAfilter Maxi cartridge, placed in a convenient tube and incubated at RT for 10 min.

The QIAfilter cartridge was then hold over the previously equilibrated QIAGEN-tip and the cap was removed from the QIAfilter outlet nozzle. The Plunger was gently inserted to the QIAfilter Maxi cartridge to filter the cell lysate into a 50 ml tube. Then 2.5 ml endotoxin removal buffer ER was added to the filtered lysate, the sample mixed by inverting the tube about 10 times and incubated on ice for 30 min.

During incubation a QIAGEN-tip 500 was equilibrated by applying 10 ml buffer QBT and the column was allowed to empty by gravity flow.

Then the incubated filtrate was applied to the QIAGEN-tip and allowed to enter the resin by gravity flow. The QUIAGEN-tip was then washed with 2 x 30 ml buffer QC. DNA was eluted with 15 ml buffer QN (into an endotoxin-free 30 ml Nalgene-tube). Plasmid DNA was precipitated with 10.5 ml isopropanol at RT and centrifuged at 11,000 rpm (Sorvall RC-5B, rotor HB-4 11,000 rpm) at 4°C for 30 min.

The pellet was briefly washed with 70 % ethanol and again centrifuged at 11,000 rpm at 4°C for 15 min. Finally, the pellet was air dried and DNA was dissolved in a suitable volume of buffer EB (10 mM Tris-HCl, pH 8.5).

Determination of DNA/RNA Concentration

The concentration and purity of nucleic acids was determined by measuring the optical density (OD) at 260 nm and 280 nm in a SmartSpec 3000 spectrometer (Biorad). An OD_{260} of 1 represents either 40 µg/ml RNA or ssDNA or 50 µg/ml dsDNA. The OD_{260}/OD_{280} ratio between 1.8 and 2 indicates high purity of nucleic acid. Contaminants that absorb at 280 nm (e.g. protein) will lower this ratio.

DNA Sequencing

Sequencing of dsDNA was performed with "ABI PRISM Big Dye Terminator v3.1 Cycle Sequencing Kit" according to manufacturer guidelines provided with the kit. The following reagents were combined:

reagent	µl per reaction tube
250-500 ng DNA diluted in aqua dest.	4.4
sequence primer (5 pmol/µl)	1.6
terminator ready reaction mix	4
total volume	10

The following amplification parameters were chosen in a "PE GeneAmp PCR System 2400" or "PE GeneAmp PCR System 9700":

1 cycle:	1 min	96°C
28 cycles:	10 sec	96°C
	15 sec	50°C
	2 min	60°C
hold at:		4°C

Samples were transferred to the "in house sequencing lab", where DNA sequencing was performed on an ABI-Prism 3100 System (Applied Biosystems).

Specific sequencing primers were chosen for each approach. As the maximal fragment length being analysed does not exceed 500-800 bp, analyses of longer sequences of interest were divided into 500 bp intersections.

Sequence data were edited and analysed with the software VectorNTI of Invitrogen.

Freezing of Bacteria

800 µl of cultured bacteria suspension was mixed with 800 µl of sterile glycerol in a cryotube and mixed to ensure that the glycerol was homogenously dispensed. The cultures were stored at -80°C.

Cell Biology

Cell Lines

All used cell lines were grown in an atmosphere containing 5 % (v/v) CO_2 and 95 % relative humidity. All cell lines except HN5 and T47D Tet-on were obtained from ATCC and were propagated according to provider's instructions.

HN5

HN5 is a human head and neck squamous cell carcinoma cell line which was originally obtained from Invitrogen, CA, USA.
For medium condition see chapter "Media and Buffers".

T47D Tet-on

T47D Tet-on cell line (BD Pharmingen) stably expressing a tet-repressor, was propagated following provider's instructions. "Tet System Approved Fetal Bovine Serum" from Clontech (Cat.# 631101) was used for cultivation of the tet-sensitive cell line. For induction, Doxycycline (Clontech, Cat# 631311) was used at a concentration of 2 µg/ml media.

Thawing and Freezing of Cells

At least $1\cdot10^6$ cells in excellent condition (derived from log growth phase) were centrifuged at 1,000 rpm for 2 min in a Megafuge 2.0R (Haereus). Afterwards the supernatant was removed. The pellet was resuspended gently in 1 ml of the pre-cooled freeze-medium "Cryo-SFM" (PromoCell, Cat.# C-29912) and transferred into a cryovial which was labelled with an ethanol-resistant marker. Then the vial was put into a freeze container (NALGENE™ 1°C Cryo Freezing Container) and transferred into a -80°C freezer, whereby the temperature drops 1°C per min avoiding ice crystal formation. If cells should be stored longer than 6 months, they should be transferred to liquid nitrogen afterwards.

For thawing, frozen cells were put immediately from dry ice into a 37°C water bath, until the freeze-medium thawed. When only a small lump of ice was left, the vial was removed from the water bath and sprayed briefly with 70 % ethanol to avoid contamination. The vial was opened in a laminar flow and the cells were immediately transferred to a T25 cell culture flask and resuspended with 5 ml of their respective growth medium. After 24 h, the cells were washed with 1x PBS and the medium was changed. After one week in culture, usually the cells are in a good condition and ready for experiments.

Propagation of Adherent Cells

The cells were trypsinised when confluency of ~ 90 % was reached.
For this, the cell culture medium was removed and cells were washed with 1x PBS. Then 1 ml (T75 flask) or 2.5 ml (T175 flask) of Trypsin-EDTA solution ("Trypsin-EDTA (1x), liquid - 0.05 % trypsin" of Invitrogen, Cat.# 25300-054) was added and the flasks incubated at RT (or at 37°C) until the cells were detached.
The cells were "split" according to their density and growth rate. Fresh culture medium was added and dispensed into new culture flasks.

Determination of Cell Count

For determination of cell count, cells were detached with Trypsin-EDTA solution as described above and the trypsinisation reaction stopped with the respective cell medium. A homogenised cell solution or dilution thereof was measured with the "Vi-CELL™ Cell Viability Analyzer" from Beckman Coulter.

Determination of Growth Curve

For determination of cell growth, in 2D cell culture, a defined number of cells were seeded into a cell culture flask or plate or well and after a certain period of time tyrpsinised and the cell solution counted with the "Vi-CELL™ Cell Viability Analyzer" from Beckman Coulter. With enough time points a growth curve could be compiled in MS Excel.
For experiments in 3D cell culture systems with spheroids, the area of spheroids was determined on different days by using the software „AxioVision Rel.4.6" from Carl Zeiss Imaging Solutions.

Transient Transfection of Plasmid DNA to Eukaryotic Cells

For transient transfection of eukaryotic cells with plasmid DNA the "Lipofectamine 2000" reagent of Invitrogen (Cat.# 11668-019) was used.

The day before transfection, cells were trypsinised and cultivated at a confluency of ~ 70 % o/n. For transfection of cells in 6-well-cell culture plates, 200,000 cells were seeded and grown in 2 ml of the appropriate media for at least one day. 2 µg plasmid per reaction was diluted with 200 µl OptiMEM (Invitrogen, Cat.# 51985-026), and 5 µl Lipofectamine 2000 was diluted with 200 µl OptiMEM. After 15 min incubation at RT both solutions were combined and incubated for further 15 min at RT. The cells were washed once with PBS and then treated with the lipofectamine-plasmid solution for 4 h in the incubator. 4 h after transfection, media was changed to the respective cell medium.

Usually cells were incubated for 24-48 h in the incubator prior to testing for transgene expression.

Generation of Stable Cell Lines

For the establishment of CLCA2 stably transfected inducible T47D Tet-on cell lines, or cell lines with distinct mutations in CLCA2, the ORF's of CLCA2 with an HA-tag at the C-terminal end and various mutated CLCA2-ORF's were cloned to the pTRE2pur vector of the "Tet-on Gene Expression System" of Clontech (Cat.# 630922) and transfected to the T47D Tet-on cell line.

In the Tet-on System, rtTA (reverse tetracycline-controlled transactivator) binds the TRE (Tet response element) and activates transcription only in the presence of Doxycycline (a Tetracycline derivative), whereby in the used T47D Tet-on cell line, the appropriate regulatory protein is already stably expressed.

CLCA2-HA-expressing stable cell clones, having the puromycin resistance gene from the transfected vector pTRE2pur, were selected for at least 2 weeks with 0.5 µg/ml Puromycin (Sigma, Cat.# P9620) in cultivation medium. Approximately 100 cell clones were analysed for Doxycycline-dependent expression of CLCA2-HA by using an anti-HA-antibody in Western blot analyses with induced versus non-induced cell lysates, which were generated 48 h post induction.

Those clones with no expression in uninduced cell lysates, but distinct expression in induced cell lysates were used for further analysis.

Generation of Cell Lysates from Human Tumour Cells

To generate lysates of adherent cells, the medium and the detached cells were removed. The remaining attached cells were washed once with PBS, and then 150 µl lysis buffer was added to each well of a 6-well-plate or 1 ml lysis buffer to cells which were grown in a 13.5 cm culture dish. The lysis buffer was pipetted up and down, until all cells were detached. Then the lysate was immediately put to the freezer and stored for at least 2 h at -20°C.

Inhibition of Post-Translational Modifications

The glycosylation inhibitor Tunicamycin and isoprenylation-inhibitor Perillic acid were obtained from Sigma. The inhibitors were dissolved in DMSO and added directly to the cell culture medium at different concentrations. Final DMSO concentration in all samples was adjusted to 1 %. 48 h after treatment cells were harvested and protein lysates tested in Western blot analysis.

Proliferation Assay (Alamar Blue)

1,000-5,000 cells in 100 µl medium were seeded per well of a 96-well-plate. Proliferation assays were performed by adding 100 µl of 20 % Alamar reagent (Invitrogen, Cat.# DAL1100) in medium into each well. After subsequent incubation at 37°C the measurements were performed in a standard microtiter well plate reader such as EnVision 2101 multi label reader (Perkin Elmer).

siRNA Experiments

siRNAs

siGENOME SMART pool (Dharmacon, Cat.# MQ-003813-01-0005, human CLCA2, NM_006536, 5nmol x 4) containing a mixture of four designed siRNAs targeting CLCA2 mRNA was tested in different cell lines for efficient downregulation of the specific mRNA and protein levels. Afterwards the individual siRNAs, shown in Table 6 were tested individually.

Name of siRNA	Sequence (5' → 3')	Dharmacon Cat.#
siCLCA2-2	GGAAUCAUUUGCCUUAUUAUU	D-003813-02
siCLCA2-4	GAUCGAAAGUUGCUGGUUUUU	D-003813-04
siCLCA2-5	UGACAAACCUUUCUACAUAUU	D-003813-05
siCLCA2-8	GCUACAAGCUAUGAAAUAAUU	D-003813-08

Table 6. Sequences of commercial available (Dharmacon) siRNAs, which were used for CLCA2 knock-down experiments

The most efficient siRNA oligonucleotides, siCLCA2-2 and siCLCA2-4, were used for further experiments.

Non-targeting siRNA "siCONTROL#1" (herein named siCTRL) of Dharmacon (Cat.# D-001210-0105) was also purchased and used as a control for normalisation.

siRNA Transfection in 6-Well-Plates

For each siRNA transfection approximately 150,000 cells were seeded into each well of a 6-well-cell culture plate and cultured for 24 h to yield 40-50 % confluency.

siRNA duplexes were dissolved and diluted with siRNA buffer (Dharmacon, Cat.# B-002000-UB) and nuclease free water to a concentration of 20 µM.

Cells were transfected with siRNAs dependent on the cell line with the indicated amount of appropriate DharmaFECT^{TM} reagent at a final concentration of 25 nM siRNA.

For this, 2.5 µl siRNA solution (20 µM) was diluted in 200 µl of a serum-fee respective cell culture medium. In parallel, 4 µl DharmaFECT was diluted in 200 µl of the serum-free respective cell culture medium. Both solutions were incubated separately. After 5 min of incubation the solutions were combined and incubated at RT for further 20 min. Then 1.6 ml of the corresponding full cell culture medium (including serum) was added to 400 µl of the mixture to obtain 2 ml transfection medium. After washing the target cells once with PBS, the medium of the target cells was replaced with 2 ml of the transfection medium. After 24 h of incubation, the transfection medium was discarded and 2 ml respective cell culture medium was added.

The cells were harvested 24-72 h after siRNA transfection and analysed via quantitative PCR or Western blot analysis.

siRNA Transfection in 96-Well-Plates

For performing an Alamar Blue Assay with siRNA-transfected cells, siRNA transfections were performed in 96-well-plates. Therefore 5,000 cell per well were seeded into each well and final siRNA concentrations ranging from 390 pM up to 100 nM were tested for optimal transfection. (In detail, the following concentrations were used: 390 pM, 780 pm, 1.56 nM, 3.13 nM, 6.25 nM, 12.5 nM, 25 nM, 50 nM, 100 nM)

100 µl transfection reagent was added to each well of a 96-well-plate.

If RNA was isolated from such an experiment, the "FastLane Cell cDNA Kit" from Qiagen (Cat.# 215011) was used following the instructions in the manual.

DharmaFECT Reagents

Table 7 lists all transfected cell lines used in this study and the respective DharmaFECT reagent, which was used for transfection.

Cell Line	DharmaFECT	Dharmacon Cat.#
HN5	4	T-2004
MDA-MB-453	1 or 2	T-2001 & T2002
T47D Tet-on	4	T-2004

Table 7. Optimal transfection reagents (Dharmacon) for cell lines used in this study

Live-Cell High-Content Screening (HCS) with Cellomics ArrayScan®

HCS Reader is an automated fluorescence microscopic imaging system designed for high content screening and high content analysis. The instrument features include optics by Carl Zeiss, broad white-light source, 12-bit cooled CCD camera, and controller software. All Cellomics instruments are designed to work with image analysis modules (BioApplications) that automatically convert images into numeric data that capture changes in cell size, shape, intensity, and other properties.

The following Standard Operating Procedures for cell cycle and apoptosis analysis (CCYAPO) have been applied:

Tissue Culture

Cells of interest were grown in maxi plates (175 cm^2) using EMEM medium supplemented with 10 % heat-inactivated fetal calf serum, 0.15 % NaHCO3, 1 % Na pyruvate solution, 1% NEAA (100x) and 2 mM L-Glutamine. Cultures were incubated at 37°C and 5 % CO_2 in a humidified atmosphere.

Cell were transfected according the standard protocol (see previous chapters).

Staining Procedure

All liquid handling was performed with Cybiwell instruments.

Apoptosis staining was performed on the life cells by adding 50 µl/well medium containing 500 nM Mitotracker Red CMXRos (1mM stock in DMSO, Invitrogen M-7512) and 100 nM SytoxGreen (100 µM stock in DMSO; Invitrogen S-7020). Then the cells were kept in the incubator for 60 minutes. After incubation the cell layer was washed 2 times with 100 µl PBS. Then cells were fixed by adding 100 µl 3.7 % formaldehyde/ 0.1 % TX100 in PBS for 10 min at RT. The cell layer was washed 2 times with 100 µl PBS. The nuclei were stained with 5 µg/ml Hoechst 33342 (stock 10 mg/ml in PBS; Invitrogen H1399) for 60 min. Then the cell layer was washed 1 x with 100 µl 0.1 % Triton X100 in PBS and 2 x with 100 µl PBS. The wells were filled with 100 µl PBS and the plates sealed with an adhesive sheet.

The measurement was performed with Cellomics "ArrayScan® VTI Live" (Protocol: Morphology Analysis TV_APO_20x) as follows:

Settings

- objective: 20x
- acquisition camera mode: standard
- filter XF93: DAPI, FITC and TRIC channels
- autofocus interval = 2
- 600 cells, max 30 fields/well

MATERIAL AND METHODS

- max 5 sparse fields per well
- fixed exposure time, set to 25 % saturation for the negative controls
- fixed threshold - adjusted to the actual staining intensities
- SegmetationCh1 = 6
- background correction = 0
- NucSmoothFactorCh1 = 0

Features

- ObjectTotalIntenCh1   Hoechst stain: DAPI channel
- ObjectAreaCh1
- ObjectVarIntenCh1
- TotalIntenCh2   Sytox stain: FITC channel
- TotalIntenCh3   MitoTracker: TRITC channel
- CellsPerField

Sytox intensities were generated by dividing – on the single cell level – the TotalIntenCh2 by TotalIntenCh1 values.

Integrin-Mediated Cell Adhesion Assays

Integrin-mediated cell adhesion assays were performed by using „Beta Integrin-Mediated Cell Adhesion Array" (Chemicon, Cat.# ECM531) and „Alpha Integrin-Mediated Cell Adhesion Array" (Chemicon, Cat.# ECM530).

The „Alpha Integrin-Mediated Cell Adhesion Array" uses mouse monoclonal antibodies, generated against human α-integrins/subunits ($\alpha 1, \alpha 2, \alpha 3, \alpha 4, \alpha 5, \alpha V$ and $\alpha_V \beta_3$) and the „Beta Integrin-Mediated Cell Adhesion Array" uses mouse monoclonal antibodies, generated against human β-integrins/subunits (β1, β2, β3, β4, β6, αVβ5, and α5β1), which are immobilised onto a goat anti-mouse antibody coated microtiter plate. The plate is used to capture cells expressing these integrines on their cell surface. Cells were seeded onto the coated substrate and incubated. Unbound cells were washed away and adherent cells were fixed and stained. Relative cell attachment was determined using absorbance readings, measured in a standard microtiter well plate reader.

The assays were performed according to the assay instructions in the manual.

For cell suspension 1.0 x 10^6 cells/ml were used.

Human Tissue Samples

Human tissues were collected at the Institute of Pathology, Medical University of Vienna, Austria, following institutional guidelines. In addition, ten cases of cervical SCCs were obtained from Oridis Biomed.

3D-*In Vitro* Carcinoma Assay: Spheroids

For investigation of CLCA2-expressing cell lines in a 3D model, cells were grown in the form of spheroids. To achieve growth in this tumour similar format, cells have to be cultivated with methylcellulose in wells of a 96-well-plate for suspension culture, where cells could not attach at the surface of the plate. When spheroids were formed, they were transferred to a gel, where growth in all directions is possible.

Methylcellulose

1.8 g methylcellulose (Sigma, Cat.# M-0512-100G, 4000 centipoises) was weighted into a 100 ml flask containing a magnetic stirrer and autoclaved (the methylcellulose powder is resistant to this procedure). The autoclaved methylcellulose was resolved in 100 ml sterile basal medium (using a magnetic stirrer) o/n at 4°C. After clearing by centrifugation at 5,000 g for 2 h at RT, a final stock solution was aliquoted, whereby only the clear highly viscous supernatant was used for the spheroid assay (about 90 - 95 % of the stock solution). For spheroid generation, 20 % of the stock solution and 80 % culture medium were used.

Spheroid Formation

Cells were trypsinised and counted. For one spheroid 250 cells were used, so 25,000 cells were isolated, centrifuged at 1,000 g for 3 min at RT and resuspended in 10 ml appropriate medium (containing serum and all needed "goodies" such as L-glutamine, Na-pyruvate etc.) with 20 % Methylcellulose stock solution. 100 µl cell suspension was added into each well of a U-shaped 96-well-plate for suspension culture from Greiner (Cat.# 650185). The cells were incubated for 48 h at 37°C in an atmosphere containing 5 % (v/v) CO_2 and 95 % relative humidity.

MATERIAL AND METHODS

Harvesting of Spheroids

The spheroids were collected by removing 50-100 µl of the medium near the deepest point of the well of the U-shaped 96-well-plate with a 200 µl Gilson/Eppendorf pipette and collected into a 1.5 ml Eppendorf tube. The collected spheroids were centrifuged carefully at 100 g for 3 min at RT and the medium was removed with a pipette.

Gel Pouring Device

For growing spheroids in collagen gel, a gel pouring devices is needed to keep the stability of the gel.

For this, silicon gaskets were cut out in quadratical (2 cm x 2 cm) pieces from silicon gasket (Biorad silicone gasket for gel dryers) and a hole (diameter 1.5 cm, with a hollow punch) was made in the middle. The gaskets were autoclaved (45 min, 120°C) before use. The supporting nylon mesh was taken of Becton Dickinson Filcons (100 µm mesh size, Cat.# 340611) and autoclaved (45 min, 120°C).

To be used as a form for 3D-spheroid-gels, the silicone gaskets were placed onto the lid of a 4 cm cell culture dish with a nylon mesh placed into the hole and incubated for 30 min at 37°C before use.

Collagen Gel Preparation

For collagen gels, "Collagen I, Rat Tail Tendon" from BD Biosciences (Cat.# 354236) was used.

The following reagents (for 2 ml collagen gel mix) were put on ice and mixed carefully by pipetting up and down for about 10 times avoiding production of bubbles:

        0.2 ml 10x PBS
        0.9 ml Methylcellulose (20 %) in desired culture medium
        0.9 ml Collagen I
        23 µl   1M NaOH

180 µl of ice cold collagen gel mix was added to desired spheroids and mixed carefully by pipetting up and down for about 10 times without producing bubbles.

The fluid spheroid gel was then pipetted into the pre-warmed pouring device with mesh and immediately put back to incubator, where the collagen hardens over a period of 30 min. Then, with the help of a sterile forceps the silicone foil was carefully remove and the collagen gel transferred into a 24-well-plate with the appropriate culture medium. Thereby the gel was only touched at the mesh supported border.

The 3D-spheroid-gel was incubated at 37°C and the growth monitored every day with the microscope.

Matrigel

As alternative to collagen, spheroids were also grown in "Growth Factor Reduced Matrigel" of BD Biosciences (Cat.# 356231).

150 µl Matrigel was used for one gel. Silicon gaskets and the nylon mesh were not used, because Matrigel very quickly hardens; so it was sufficient to form a drop and transfer the gel pouring device immediately to the 37°C incubator.

Probidium Iodide Staining

To observe the status of apoptotic cells in spheroids within a spheroid-gel, probidium iodide was added directly to the media (ratio 1:4000). This treatment easily allows observation with a fluorescence-microscope utilising the appropriate filter, exhibiting apoptotic cells as red points.

Measuring of Diameter and Volume of Spheroids

Photographs of the spheroids were taken every two days after embedding spheroids into the gel, with the microscope. The area of the spheroids were determined with the software „AxioVision Rel.4.6" from Carl Zeiss Imaging Solutions. Growth curves were calculated with MS Excel.

Biochemistry

Determination of Protein Concentration in Cell Lysates (Bradford Assay)

To normalise the different protein concentration for quantitative analysis of Western blots, first the protein concentration of protein lysates were analysed in a Bradford assay.

Protein lysates were pre-cleared by centrifugation at 14,000 rpm for 10 min at 4°C. The amount of whole protein content was measured by a standard Bradford assay, purchased of Biorad (Cat.# 500-0006). The sample was measured in parallel to samples of a calibration curve, consisting of samples with defined amounts of BSA (0 / 1 / 5 / 7 / 10 / 15 µg/ml BSA). 5 µl of each protein sample was mixed with 1 ml 1x Bradford solution and incubated for 5 min at RT. The optical density of each sample was measured at a wavelength of 595 nm in a SmartSpec 3000 spectrophotometer. The gradient of the calibration curve was calculated and used to assess the protein concentration of the samples.

SDS-PAGE

All SDS-PAGE analyses were performed using pre-cast Criterion XT Bis-Tris, 4-12 % gradient gels, 18-well, 30 µl, acquired from Biorad (Cat.# 345-0124), using the Biorad gel chamber for Criterion gels.

Depending on different approaches, 50-100 µg total protein was used for SDS-PAGE analysis. The specified amount of protein was prepared with 5x SDS-LB and then boiled at 95°C for 5 min. The sample was cooled for 1 min at ice and spun down prior loading onto the protein gel. For determination of size of proteins, as marker was used "PageRuler Prestained Protein Ladder" of Fermentas (Cat.# SM0671).

The electrophoresis was started at 100 V and subsequently elevated up to 140 V.

Western Blotting

The transfer of electrophoretically separated proteins to Immuno-blot PVDF membranes (Biorad, Cat.# 162-0239) was performed by Semidry-blotting a TE 70 Semi-Dry Transfer Unit at 170 mA for 1 h per membrane.

The membrane was equilibrated for a few seconds in methanol (MeOH) prior use. For transfer the membrane and the gel were incubated and the filter papers soaked in 1x transfer buffer. After the transfer, the blots were blocked in blocking solution (TBS-T plus 5 % dried milk) for 1 h at RT and then incubated with the primary antibody diluted in blocking solution o/n at 4°C. The next day, the antibody dilution was discarded and the membrane was washed 3 x for 10 min with TBS-T. Then the membrane was incubated with the respective horse-radish-peroxidase-conjugated secondary antibody, diluted in blocking solution, for 1 h at RT. Afterwards the membrane was again washed 3 x for 10 min with TBS-T.

Detection of bound antibodies was performed by chemiluminescence, using "Amersham ECL Western blotting detection reagents" from GE Healthcare (Cat.# RPN2106) and exposing the membrane to a light-sensitive film ("Amersham Hyperfilm MP" of GE Healthcare, Cat.# RPN1675K) for an appropriate time in an exposure-cassette. The film was developed using a Curix developer.

Proteome Analyses with Proteome Profiler™ (R&D Systems)

MAPK-Array

For analysing the phosphorylation status of all three major families of mitogen-activated protein kinases (MAPKs), the extracellular signal-regulated kinases (ERK1/2), c-Jun N-terminal kinases (JNK1-3), different p38 isoforms ($\alpha/\beta/\delta/\gamma$), other intracellular kinases such as Akt, GSK-3 and p70 S6 kinases, the "Human Phospho-MAPK Array Kit" from R&D Systems (Cat.# ARY002) was used.

The principle of the assays is that capture and control antibodies have been spotted in duplicate on nitrocellulose membranes. Cell lysates are diluted and incubated with the Human Phospho-MAPK Array. After binding both, phosphorylated and non-phosphorylated kinases, unbound material is washed away. A cocktail of phospho-site-specific biotinylated antibodies is then used to detect phosphorylated kinases via Streptavidin-HRP and chemiluminescence.

RTK-Array

For analysing the relative levels of phosphorylation of 42 different receptor tyrosine kinases, the "Human Phospho-RTK Array Kit" from R&D Systems (Cat.# ARY001) was used.

The principle of the assays is that capture and control antibodies have been spotted in duplicate on nitrocellulose membranes. Cell lysates are diluted and incubated with the Human Phospho-RTK Array. After binding the extracellular domain of both, phosphorylated and non-phosphorylated RTKs, unbound material is washed away. A pan anti-phospho-tyrosine-antibody conjugated to horseradish peroxidase (HRP) is then used to detect phosphorylated tyrosines on activated receptors by chemiluminescence.

Immunohistochemistry

An indirect immunoperoxidase assay was used to detect expression of CLCA2. 5 µm thick sections were cut and mounted on poly-(L-lysine)-coated slides. After de-paraffinisation and epitope retrieval with Proteinase K in TrisEDTA buffer pH 8.0 for 15 min at 37°C, the sections were blocked with 10 % normal goat serum. The primary antibody was applied for 1 h at RT followed by incubation with goat anti-rabbit horseradish peroxidase labelled polymer, Dako Envision System (Dako) for 30 min at RT. The final reaction product was visualised with DAB (0.06 % 3,3'-diaminobenzidine in PBS, 0.003 % H_2O_2) for 2-5 min. Then the slides were dehydrated in alcohol and counterstained with hematoxylin. The specificity of the staining was confirmed by pre-incubating the primary anti-peptide-antibody with the corresponding synthetic peptide for 2 h at RT at the final concentration of 25 µg/ml of antibody which resulted in no staining of the tumour cells.

Antibody Generation

CLCA2 antibodies were generated by immunising rabbits with peptides corresponding to the amino acid sequence of CLCA2 (NH2-epitope-COOH). Three rabbits were immunised with the following peptides: Table 8

Rabbit No.	Amino Acid Pos. in CLCA2 Protein Sequence	Peptide Sequence
1	K94 – N110	KANNNSKIKQESYEKAN
	N179 – D199	NDKPFYINGQNQIKVTRCSSD
2	E498 – T513	ESTGENVKPHHQLKNT
	H690 – H708	HVNHSPSISTPAHSIPGSH
3	T643 – L656	TVEPETGDPVTLRL

Table 8. Peptides used for generation of antisera in rabbits

The serum containing the anti-CLCA2-antibodies was affinity purified using the corresponding peptide coupled to a Sulfo-Link-Gel-column, using the SulfoLink Kit from Pierce (Cat.# 44895). The antiserum was eluted with 0.2 mol/l Glycine/HCl pH 2.7, 150 mmol/l NaCl. The fractions were tested in Western blotting analysis.

In Silicio Analyses

Statistical and Bioinformatic Analyses

All mathematical calculations and graphics were performed with MS Excel and GraphPad Prism 4.0.

Sequence Alignment

Sequence data were edited and analysed with the VectorNTI Software of Invitrogen.

Expression Profiling

For expression analysis, box-and-whisker plots were generated as described [75;168]. The respective hybridisations were performed on Affymetrix Exon GeneChips. These microarrays are based on 25-mer oligonucleotides and allow the detection of more than 39,000 human transcripts, with probe sets of 11 oligonucleotides used per transcript.

Approximately 1 μg of total RNA was converted into DNA, which was fragmented and end-labeled. 5 μg of labelled target DNA was hybridised to the Affymetrix GeneChip-Human Exon 1.0 Array at 45°C for 16 h as recommended by the manufacturer (http://www.affymetrix.com). After hybridisation arrays were washed and scanned on a GCS3000Scanner (Affymetrix, Santa Clara, CA). Chip data were normalised with the statistical algorithm implemented in the Microarray Suite version 5.0 (Affymetrix).

The raw expression intensity for a given chip experiment is multiplied by a global scaling factor to allow comparisons between chips. This factor is calculated by removing the highest 2 % and the lowest 2 % of the values of the non-normalised expression values, and calculating the mean for the remaining values, as trimmed mean. One hundred divided by the trimmed mean gives the scaling factor, where 100 is the standard value used by GeneLogic. For oligonucleotide-specific background normalisation individual absent- and present-calls are calculated by comparing the absolute signal intensities for corresponding perfect-match and mismatch oligonucleotides.

Media and Buffers (in alphabetical order)

Molecularbiology

1st Strand buffer, 5x:
(Invitrogen; Cat.# 18064-014)

250 mM Tris-HCl (pH 8.3)
375 mM KCl
15 mM $MgCl_2$

Ammonium acetate, 7.5 M:

57.86 g NH_4Ac
add 100 ml H_2O

Ampicillin stock:

50 mg/ml Ampicillin in H_2O
filter through 0.22 µm sterile

Blocking solution:
(for Western blot)

1 x TBS-T buffer
5 % dried milk

Buffer EB:
(Qiagen; Cat.# 27106)

10 mM Tris-HCl
pH 8.5

Buffer P1:
(Qiagen; Cat.# 27106)

50 mM Tris-HCl (pH 8.0)
10 mM EDTA
100 µg/ml RNAseA

Buffer P2:
(Qiagen; Cat.# 27106)

200 mM NaOH
1 % SDS

Buffer P3:
(Qiagen; Cat.# 27106)

2.55 M KAc (pH 4.8)

Buffer QBT:
(Qiagen; Cat.# 12163)

750 mM NaCl
50 mM MOPS (pH 7.0)
15 % (v/v) Isopropanol
0.15 % (v/v) Triton X-100

Buffer QC: *(Qiagen;* *Cat.# 12163)*	1 M 50 mM 15 % (v/v)	NaCl MOPS (pH 7.0) Isopropanol
Buffer QF: *(Qiagen;* *Cat.# 12163)*	1,25 M 50 mM 15 % (v/v)	NaCl Tris-HCl (pH 8.5) Isopropanol
DEPC-H_2O:	0.1 % (v/v)	DEPC in Millipore water; shake well autoclave the next day 20 min at 1.5 bar
DNA loading buffer 6x:	0.3 % 0.3 % 1 % 30 % 30 mM	Bromphenole Blue Xylene Cyanol SDS Glycerol EDTA
Klenow 10x buffer: *(Promega;* *Cat.# M195A)*	500 mM 100 mM 1 mM	Tris-HCl (pH 7.2) $MgSO_4$ DTT
LB media: (Lauria-Bertani Broth) *(Biolab;* *Cat.# LBB30500)*	10 g/l 5 g/l 10 g/l	Tryptone Yeast extract NaCl pH 7.5 (approx.)
LB-Amp media:	1000 ml 1 ml	LB media 1000x Ampicillin Stock
LB-Amp plates:	1000 ml 15 g 1 ml	LB media Bacto Agar *(Difco)* autoclave 20 min at 1.5 bar 1000x Ampicillin Stock (added at 55°C)
Ligase buffer, 10x: *(Promega;* *Cat.# C126A, C126B)*	300 mM 100 mM 100 mM 10 mM	Tris-HCl (pH 7.8) $MgCl_2$ DTT ATP
PCR buffer, 10x: *(TaKaRa;* *Cat.# R001A)*	100 mM 500 mM 15 mM	Tris-HCl (pH 8.3) KCl $MgCl_2$

PCR buffer, 10x: Mg^{2+} free: *(TaKaRa; Cat.# R001AM)*	100 mM 500 mM	Tris-HCl (pH 8.3) KCl
SDS-loading buffer, 1x:	50 mM 100 mM 2 % 0.1 % 10 %	Tris-HCl (pH 6.8) DTT or β-Mercaptoethanol SDS Bromphenol Blue Glycerol
SOC media: *(Invitrogen; Cat.# 15544-034)*	2 % 0.5 % 10 mM 2.5 mM 10 mM 10 mM 20 mM	Tryptone Yeast extract NaCl KCl MgCl$_2$ MgSO$_4$ Glucose
Sodium acetate, 3M:	40.8g 80 ml add 100 ml	NaAc·3H$_2$O H$_2$O adjust with Acetic acid to pH 5.2 H$_2$O
TBE buffer, 10x: *(Invitrogen; Cat.# 15581-044)*	1 M 0.9 M 0.01 M	Tris Boric acid EDTA
TBS-T: (Tris buffered saline with 0.05 % Tween 20)	20 mM 150 mM 0.05 %	Tris-HCl NaCl Tween 20
TE:	10 mM 1 mM	Tris-HCl (pH 7.5 - 8) EDTA
Transfer buffer:	20 mM 66.6 mM 20 %	Tris-HCl Glycine Methanol

Cell Culture

PBS, Dulbecco's (1x):	0.2 g/l	KCl
(Gibco; Cat.# 14190-094)	0.2 g/l	KH_2PO_4
	8 g/l	NaCl
	1.15 g/l	Na_2HPO_4
		pH 7.2

Lysis buffer for protein lysates:	20 mM	Hepes, pH 7.3
	5 mM	EDTA
	2 nM	EGTA
	30 mM	Sodiumfluoride
	1 mM	Sodiumorthovanadate (Na_3VO_4)
	20 mM	Sodiumpyrophosphate
	40 mM	b Glycerophosphate
	30 mM	Sodiumfluoride
	0.5 %	Triton X-100
	1 Tab. / 50 ml buffer	Protease Inhibitor Cocktail, "complete, EDTA-free", Roche, Cat.# 04693132001

Mammalian cell culture media for T47D Tet-on cell line:	~450 ml	DMEM high glucose medium, with L-glutamine
	10 %	Fetal bovine serum, Tet System approved (Clontech, Cat.# 631101)
	5 ml	100x Hepes buffer
	2 ml	Geneticin (50mg/ml)
	5 ml	100x Sodium pyruvate
	0.5 ml	Bovine Insulin (10g/ml)

Mammalian cell culture media for HN5 cell line:	~450 ml	DMEM high glucose medium, with L-glutamine
	10 %	Fetal Bovine Serum
	5 ml	100x Hepes buffer
	5 ml	100x Sodium pyruvate
	5 ml	100x NEAA Mixture

Mammalian cell culture media for MDA-MB-453 cell cine:	450 ml	DMEM high glucose medium, with L-glutamine
	10 %	Fetal Bovine Serum (GIBCOBRL)

Reagents for cell culture media: (in alphabetical order)
- Bovine Insulin,10 mg/ml insuline in 25 mM HEPES, pH 8.2, Sigma, Cat.# I0516
- DMEM high glucose medium, with L-glutamine, Lonza Group, Cat.# BE12-604F
- Doxycycline 5 g, Clontech, Cat.# 631311
- D-PBS (1x), Invitrogen, Cat.# 14190-094
- Fetal bovine serum, JRH Biosciences, Cat.# 12103-500M
- Geneticin selective antibiotic, Invitrogen, 50 mg/ml Cat.# 10131-019
- GlutaMAX-I supplement, 200 mM, Invitrogen, Cat.# 35050-038
- Hepes buffer 1M, Cambrex, Cat.# BE17-737E
- MEM medium, Invitrogen, Cat.# 21090-022
- MEM non essential amino acids solution (100x), Invitrogen, Cat.# 11140-035
- Puromycin, Sigma, Cat.# P9620
- RPMI 1640 medium w/GlutaMAX, Invitrogen, Cat.# 61870-036
- Sodium Pyruvate MEM 100 mM, Invitrogen, Cat.# 11360-039
- Tet system approved fetal bovine serum, Clontech, Cat.# 631101

Primer

All oligo-deoxynucleotides were ordered from metabion GmbH. The crystals were dissolved in elution buffer EB (10 mM Tris-HCl, pH 8.5) to a concentration of 100 pmol/µl. Aliquotes of 5 pmol/µl were prepared.

Cloning Primer

T_m (in PCR buffer) = 22 + 1,46 (2*(G+C)+A+T)

Name	EBI-No.	Sequenz (5' → 3')
hCLCA2-EcoRI f	15911	ccggaattcatgacccaaaggagcattgc
hCLCA2-XhoI r	15912	ccgctcgagttataataattttgttccattctctttc
CLCA2-BamHI r	15957	cgggatcctaataattttgttccattctctttcttgt
hCLCA2-NotI f	15958	ataagaatgcggccgccaccatgacccaaaggagcattgc

Table 9. Sequences of cloning primer

MATERIAL AND METHODS

Cloning Primer for *In Vitro* Site-Directed Mutagenesis of CLCA2

Primers were designed according provider's instructions, the melting temperature was calculated according provider's Website:
http://www.stratagene.com/QPCR/tmCalc.aspx

Mutation	Basepaire Change	EBI-No.	Sequence of Primer (5' → 3')
C132A	TGT → GCC	17201 (f)	ccctacaatacagagggGCCggaaaagagggaaaatacattcatttcac
		17202 (r)	gtgaaatgaatgtatttccctcttttccGGCccctctgtattgtaggg
C196A	TGT → GCC	- (f)	caaattaaagtgacaaggGCCtcatctgacatcacaggc
		- (r)	gcctgtgatgtcagatgaGGCccttgtcactttaatttg
C206A	TGT → GCC	- (f)	cacaggcattttgtgGCCgaaaaaggtccttgcc
		- (r)	ggcaaggacctttttcGGCcacaaaaatgcctgtg
C211A	TGC → GCC	- (f)	gtgtgtgaaaaaggtcctGCCccccaagaaaactg
		- (r)	cagttttcttggggGGCaggacctttttcacacac
C216A	TGT → GCC	- (f)	ggtccttgcccccaagaaaacGCCattattagtaagc
		- (r)	gcttactaataatGGCgttttcttgggggcaaggacc
N150A	AAC → GCC	16863 (f)	cctaatttcctactgaatgatGCcttaacagctggctacggatcacg
		16864 (r)	cgtgatccgtagccagctgttaagGCatcattcagtaggaaattagg
N522A	AAC → GCC	16865 (f)	ggataatactgtgggcGCcgacactatgtttctagttacgtggc
		16866 (r)	gccacgtaactagaaacatagtgtcGGcgcccacagtattatcc
N822A	AAT → GCC	16867 (f)	gctattttagtaGCCacatcaaagcgaaatcctcagcaagctggcatcaggg
		16868 (r)	ccctgatgccagcttgctgaggatttcgctttgatgtGGCtactaaaatagc
C132S	TGT → TCC	17199 (f)	ccctacaatacagagggtCCggaaaagagggaaaatacattcatttcac
		17200 (r)	gtgaaatgaatgtatttccctcttttccGGacctctgtattgtaggg

MATERIAL AND METHODS

H164C	CAT → TGC	17207 (f)	gaggccgagtgtttgtcTGCgaatgggcccacctc
		17208 (r)	gaggtgggcccattcGCAgacaaacactcggcctc
H164N	CAT → AAC	16968 (f)	gaggccgagtgtttgtcAaCgaatgggcccacctc
		16969 (r)	gaggtgggcccattcGtAgacaaacactcggcctc
H164Y	CAT → TAC	16970 (f)	gaggccgagtgtttgtcTaCgaatgggcccacctc
		16971 (r)	gaggtgggcccattcGtAgacaaacactcggcctc
E165D	GAA → GAC	17203 (f)	cgagtgtttgtccatgaCtgggcccacctccgttg
		17204 (r)	caacggaggtgggcccaGtcatggacaaacactcg
E165Q	GAA → CAG	16966 (f)	cgagtgtttgtccatCaGtgggcccacctccgttg
		16967 (r)	caacggaggtgggcccaCtGatggacaaacactcg
H168C	CAC → TGC	17209 (f)	catgaatgggccTGcctccgttggggtgtg
		17210 (r)	cacaccccaacggaggCAggcccattcatg
H168N	CAC → AAC	16972 (f)	catgaatgggccAacctccgttggggtgtg
		16973 (r)	cacaccccaacggaggtTggcccattcatg
H168Y	CAC → TAC	16974 (f)	catgaatgggccTacctccgttggggtgtg
		16975 (r)	cacaccccaacggaggtaggcccattcatg
N186A	AAT → GCC	17205 (f)	caatgacaaaccttctacataGCCgggcaaaatcaaattaaagtgacaaggtgttc
		17206 (r)	gaacaccttgtcacttttaatttgattttgcccGGCtatgtagaaaggtttgtcattg
G670A	GGA → GCC	17211 (f)	gctgatgttataaaaaatgatgCCatttactcgaggtattttttctcctttgctgc
		17212 (r)	gcagcaaaggagaaaaaatacctcgagtaaatGGcatcattttttataacatcagc
I671A	ATT → GCC	17213 (f)	gctgatgttataaaaaatgatggaGCCtactcgaggtattttttctcctttgctgc
		17214 (r)	gcagcaaaggagaaaaaatacctcgagtaGGCtccatcattttttataacatcagc
I671L	ATT → CTG	17386 (f)	gctgatgttataaaaaatgatggaCtGtactcgaggtattttttctcctttgctgc
		17387 (r)	gcagcaaaggagaaaaaatacctcgagtaCaGtccatcattttttataacatcagc
Y672A	TAC → GCC	17215 (f)	gctgatgttataaaaaatgatggaattGCctcgaggtattttttctcctttgctgc
		17216 (r)	gcagcaaaggagaaaaaatacctcgagGCaattccatcattttttataacatcagc
Y672E	TAC → GAG	17380 (f)	gctgatgttataaaaaatgatggaattGaGtcgaggtattttttctcctttgctgc
		17381 (r)	gcagcaaaggagaaaaaatacctcgaCtCaattccatcattttttataacatcagc
Y672F	TAC → TTC	17384 (f)	gctgatgttataaaaaatgatggaatttTctcgaggtattttttctcctttgctgc
		17385 (r)	gcagcaaaggagaaaaaatacctcgagAaaattccatcattttttataacatcagc
Y672K	TAC → AAG	17378 (f)	gctgatgttataaaaaatgatggaattAaGtcgaggtattttttctcctttgctgc
		17379 (r)	gcagcaaaggagaaaaaatacctcgaCtTaattccatcattttttataacatcagc
Y672Q	TAC → CAG	17382 (f)	gctgatgttataaaaaatgatggaattCaGtcgaggtattttttctcctttgctgc
		17383 (r)	gcagcaaaggagaaaaaatacctcgaCtGaattccatcattttttataacatcagc
S673A	TCG → GCC	17217 (f)	gctgatgttataaaaaatgatggaatttacGcCaggtattttttctcctttgctgc
		17218 (r)	gcagcaaaggagaaaaaatacctGgCgtaaattccatcattttttataacatcagc
R674A	AGG → GCC	17219 (f)	gatggaatttactcgGCCtattttttctcctttgctgcaaatggtagatatagc
		17220 (r)	gctatatctaccatttgcagcaaaggagaaaaaataGGCcgagtaaattccatc
R674E	AGG → GAG	17027 (f)	gatggaatttactcgGAgtattttttctcctttgctgcaaatggtagatatagc
		17023 (r)	gctatatctaccatttgcagcaaaggagaaaaaatacTCcgagtaaattccatc
R674F	AGG → TTC	17029 (f)	gatggaatttactcgTTCtattttttctcctttgctgcaaatggtagatatagcttg
		17030 (r)	caagctatatctaccatttgcagcaaaggagaaaaaataGAAcgagtaaattccatc
R674I	AGG → ATC	17025 (f)	gatggaatttactcgaTCtattttttctcctttgctgcaaatggtagatatagc
		17026 (r)	gctatatctaccatttgcagcaaaggagaaaaaataGAtcgagtaaattccatc
R674K	AGG → AAG	17362 (f)	gatggaatttactcgaAgtattttttctcctttgctgcaaatggtagatatagc
		17363 (r)	gctatatctaccatttgcagcaaaggagaaaaaatacTtcgagtaaattccatc
R674Q	AGG → CAG	17366 (f)	gatggaatttactcgCagtattttttctcctttgctgcaaatggtagatatagc
		17367 (r)	gctatatctaccatttgcagcaaaggagaaaaaataCtGcgagtaaattccatc
R674S	AGG → TCC	16545 (f)	gatggaatttactcgTCCtattttttctcctttgctgc
		16546 (r)	gcagcaaaggagaaaaaataggacgagtaaattccatc
R674T	AGG → ACC	17368 (f)	gatggaatttactcgaCCtattttttctcctttgctgcaaatggtagatatagc
		17369 (r)	gctatatctaccatttgcagcaaaggagaaaaaataGGacgagtaaattccatc

MATERIAL AND METHODS

R674Y	AGG → TAC	17364 (f)	gatggaatttactcgTaCtattttttctcctttgctgcaaatggtagatatagc
		17365 (r)	gctatatctaccatttgcagcaaaggagaaaaaataGtAcgagtaaattccatc
Y675A	TAT → GCC	17221 (f)	gatggaatttactcgaggGCCttttttctcctttgctgcaaatggtagatatagc
		17222 (r)	gctatatctaccatttgcagcaaaggagaaaaaGGCcctcgagtaaattccatc
Y675E	TAT → GAG	17372 (f)	gatggaatttactcgaggGaGttttctcctttgctgcaaatggtagatatagc
		17373 (r)	gctatatctaccatttgcagcaaaggagaaaaaCtCcctcgagtaaattccatc
Y675F	TAT → TTC	17376 (f)	gatggaatttactcgaggtTCttttctcctttgctgcaaatggtagatatagc
		17377 (r)	gctatatctaccatttgcagcaaaggagaaaaaGAacctcgagtaaattccatc
Y675K	TAT → AAG	17370 (f)	gatggaatttactcgaggAaGttttctcctttgctgcaaatggtagatatagc
		17371 (r)	gctatatctaccatttgcagcaaaggagaaaaaCtTcctcgagtaaattccatc
Y675Q	TAT → CAG	17374 (f)	gatggaatttactcgaggCaGttttctcctttgctgcaaatggtagatatagc
		17375 (r)	gctatatctaccatttgcagcaaaggagaaaaaCtGcctcgagtaaattccatc
F676A	TTT → GCC	17223 (f)	gatggaatttactcgaggtatGCCttcctttgctgcaaatggtagatatagc
		17224 (r)	gctatatctaccatttgcagcaaaggagaaGGCatacctcgagtaaattccatc
F676W	TTT → TGG	17390 (f)	gatggaatttactcgaggtattGGttctcctttgctgcaaatggtagatatagc
		17391 (r)	gctatatctaccatttgcagcaaaggagaaCCaatacctcgagtaaattccatc
F676Y	TTT → TAC	17388 (f)	gatggaatttactcgaggtattACttctcctttgctgcaaatggtagatatagc
		17389 (r)	gctatatctaccatttgcagcaaaggagaaGTaatacctcgagtaaattccatc
F677A	TTC → GCC	17225 (f)	gatggaatttactcgaggtattttGCctcctttgctgcaaatggtagatatagc
		17226 (r)	gctatatctaccatttgcagcaaaggagGCaaaatacctcgagtaaattccatc
S678A	TCC → GCC	17227 (f)	gatggaatttactcgaggtattttttcGcctttgctgcaaatggtagatatagc
		17228 (r)	gctatatctaccatttgcagcaaaggCgaaaaaatacctcgagtaaattccatc

Table 10. Sequences of mutagenesis primer

Sequencing Primers

T_m (in PCR buffer) = 22 + 1,46 (2*(G+C)+A+T)

Name	EBI-No.	Sequenz (5' → 3')
T7f	5567	taatacgactcactataggg
pUC/M13r	6298	caggaaacagctatgac
T3r	13174	attaaccctcactaaaggg
pCMV-Tag1-T3 f	7148	aataaccctcactaaaggg
pCMV-Tag1-T7 r	13432	gtaatacgactcactatagggc
pCMV6-XL4 f (VP1.5)	15909	ggactttccaaaatgtcg
pCMV6-XL4 r (XL39)	15910	attaggacaaggctggtggg
pTRE2pur f	-	cgcctggagacgccatcc
pTRE2pur r	-	cattctaaacaacaccctg
CLCA2do3	-	gcacatggagatgatccatacaccta
CLCA2do4	-	gcaagtacccacaaccaagaagcacca
CLCA2do6	-	ggaacagctaagcctgggcactggact
CLCA2do7	15906	ggttacacagcaaacggtaatattcaga
CLCA2do12	10071	gcacccaaaatgcaactgcatcaataatgt
CLCA2do13	10072	gtgatgatattagtgaccagccggaga
CLCA2up2	9626	cactttcgctcctcctcatttctgcct
CLCA2up3	9657	cctgagctgactcggctaaagcccca
CLCA2up10	9778	gacattttcacctgtactttcaagct
CLCA2up12	9818	ggtgcttcttggttgtgggtacttgca
CLCA2up13	9819	gattacatcccatgcacttctgaggct
CLCA2up19	-	gttatcattcagtaggaaattaggtgtga
CLCA2up21	-	gcacatggagatgatccatacaccta
hClCa2-check-rev	15849	gggagatatttacgttctcacc

MATERIAL AND METHODS

8F5-check-rev	15411	ggattcaatctcgtctcagc
HA-check-rev	15842	gcgtaatctggaacatcgtatg
c-myc-check-rev	15843	gtcctcctctgagatcagc
lumio-check-rev	15844	caacagccaggacaacagg
V5-check-rev	15845	ggagagggttagggatagg
6x-His-check-rev	15846	agtgatggtgatggtgatgg

Table 11. Sequences of sequencing primer

Oligomers

Name	EBI-No.	Sequenz (5' → 3')
8F5-1	15829	gatccggcagggaagttaaagctgagacgagattgaatccttagtgacccggggatatca
8F5-2	15830	agcttgatatccccgggtcactaaggattcaatctcgtctcagctttaacttccctgccg
HA-1	15831	gatcctacccatacgatgttccagattacgcttagtgacccggggatatca
HA-2	15832	agcttgatatccccgggtcactaagcgtaatctggaacatcgtatgggtag
c-myc-1	15833	gatccgagcagaagctgatctcagaggaggacctgtagtgacccggggatatca
c-myc-2	15834	agcttgatatccccgggtcactacaggtcctcctctgagatcagcttctgctcg
V5-1	15837	gatccgggaagcctatccctaaccctctcctcggtctcgattctacgtagtgacccggggatatca
V5-2	15838	agcttgatatccccgggtcactacgtagaatcgagaccgaggagagggttagggataggcttccccg
6xHis-1	15839	gatcccatcaccatcaccatcactagtgacccggggatatca
6xHis-2	15840	agcttgatatccccgggtcactagtgatggtgatggtgatgg

Table 12. Sequences of oligomers

Plasmids

pTRE2pur Expression Vector
Tet-on Gene Expression System" of Clontech, Cat.# 630922

pGEM-T Easy Vector
Promega, Cat.# A1380

pCMV-Tag1 Expression Vector
Stratagene, Cat.# 211170

Human Full-Length cDNA Clone of CLCA2 in pCMV6-XL4
Origene, Cat.# TC116023, Acc. NM_006536: Human calcium-activated chloride channel, family member 2 (CLCA2)

Antibodies

Target Protein Name	Phospho-Site	Company	Cat. No.
GAPDH (6C5)	pan	Abcam	ab8245
ß-Actin	pan	Sigma	A5441
HA-Tag (6E2)	pan	Cell Signaling	2367
V5-Tag	pan	Invitrogen	R960-25
Integrin β4 (M126)	pan	Abcam	ab29042
PARP	pan	Cell Signalling	9542
Akt	S473	Cell Signaling	#9271
Akt	S308	Cell Signaling	#9275
Akt	Y326	Cell Signaling	#2968
Akt	pan	Cell Signaling	#9272
GSK-3beta	S9	Cell Signaling	#9336
PTEN	S380	Cell Signaling	#9551
PDK1	S241	Cell Signaling	#3061
PDK1	pan	Cell Signaling	#3062
PI3K	p85 Y458 / p55 Y199	Cell Signaling	#4228
FAK	Y861	Biosource	44-626G
iKKalpha/beta	S176/180	Cell Signaling	#2697S
NF-KB p65	S536	Cell Signaling	#3031
NF-KB p65	pan	Cell Signaling	#3034
MEK 1/2	S217/221	Cell Signaling	#9121
p44/42 MAP Kinase	T202/Y204	Cell Signaling	#9101
p38 MAPK	T180/Y182	Cell Signaling	#9211
p90RSK	S380	Cell Signaling	#9341
RSK1	S221/S227	Biosource	44-924G
RSK1	S363/S369	Biosource	44-926G
Elk1	S383	Cell Signaling	#9181
Raf	S259	Cell Signaling	#9421
c-Raf	S338	Biosource	#9427
ERK 1 & 2	T185/Y187	Biosource	44-680G
Erk 1&2	pan	Biosource	44-654G
p70 S6 Kinase	T389	Cell Signaling	#9234S
p70 S6 Kinase	pan	Cell Signaling	#9202
mTOR	S2448	Cell Signaling	#2971
mTOR	S2481	Cell Signaling	#2974
4E-BP1	T37/46	Cell Signaling	#2855
4E-BP1	T70	Cell Signaling	#9455
4E-BP1	S65	Cell Signaling	#9451
FoxO1	S319	Cell Signaling	#2487
FoxO3a	S318/321	Cell Signaling	#9465
FoxO3a	S253	Cell Signaling	#9466
Stat1	Y701	Cell Signaling	#9171
Stat2	Y690	Cell Signaling	#4471
Stat3	Y705	Cell Signaling	#9131
Stat3	S727	Cell Signaling	#9134
Stat3	Y705	Cell Signaling	#9131

Target Protein Name	Phospho-Site	Company	Cat. No.
Stat5	Y694	Cell Signaling	#9351
Stat6	Y641	Cell Signaling	#9361
NPM/B23 (Nucleophosmin 1)	T234/T237	BioLegend	#619101
PKCe	pan	Biosource	AHO0743
p27	pan	novocastra	NCL-p27
Histon H3	S10	Upstate	#06-570
TSC2	T1462	Cell Signaling	#3611
TSC2	S1254	Cell Signaling	#3616
KI67	pan	Dako Cytomation	M7240
rabbit IgG-HRP conj.	pan	Sigma Aldrich	A8275
mouse IgG-HRP conj.	pan	Sigma Aldrich	A0168

Table 13. Antibodies used for Western blots for this study

APPENDIX

Sequence of CLCA2

DNA Sequence

Coding Sequence of human calcium-activated chloride channel, family member 2 (CLCA2) mRNA. ACCESSION NM_006536:

```
atgacccaaa ggagcattgc aggtcctatt tgcaacctga agtttgtgac tctcctggtt    60
gccttaagtt cagaactccc attcctggga gctggagtac agcttcaaga caatgggtat   120
aatggattgc tcattgcaat taatcctcag gtacctgaga atcagaacct catctcaaac   180
attaaggaaa tgataactga agcttcattt tacctattta atgctaccaa gagaagagta   240
tttttcagaa atataaagat tttaatacct gccacatgga aagctaataa taacagcaaa   300
ataaaacaag aatcatatga aaaggcaaat gtcatagtga ctgactggta tggggcacat   360
ggagatgatc catacaccct acaatacaga gggtgtggaa aagagggaaa atacattcat   420
ttcacaccta atttcctact gaatgataac ttaacagctg gctacggatc acgaggccga   480
gtgtttgtcc atgaatgggc ccacctccgt tggggtgtgt tcgatgagta taacaatgac   540
aaacctttct acataaatgg gcaaaatcaa attaaagtga caggtgttc atctgacatc   600
acaggcattt ttgtgtgtga aaaaggtcct tgcccccaag aaaactgtat tattagtaag   660
cttttttaaag aaggatgcac ctttatctac aatagcaccc aaaatgcaac tgcatcaata   720
atgttcatgc aaagtttatc ttctgtggtt gaattttgta atgcaagtac ccacaaccaa   780
gaagcaccaa acctacagaa ccagatgtgc agcctcagaa gtgcatggga tgtaatcaca   840
gactctgctg actttcacca cagctttccc atgaatggga ctgagcttcc acctcctccc   900
acattctcgc ttgtacaggc tggtgacaaa gtggtctgtt tagtgctgga tgtgtccagc   960
aagatggcag aggctgacag actccttcaa ctacaacaag ccgcagaatt ttatttgatg  1020
cagattgttg aaattcatac cttcgtgggc attgccagtt tcgacagcaa aggagagatc  1080
agagcccagc tacaccaaat taacagcaat gatgatcgaa agttgctggt ttcatatctg  1140
cccaccactg tatcagctaa aacagacatc agcatttgtt cagggcttaa gaaaggattt  1200
gaggtggttg aaaaactgaa tggaaaagct tatggctctg tgatgatatt agtgaccagc  1260
ggagatgata agcttcttgg caattgctta cccactgtgc tcagcagtgg ttcaacaatt  1320
cactccattg ccctgggttc atctgcagcc ccaaatctgg aggaattatc acgtcttaca  1380
ggaggtttaa agttctttgt tccagatata tcaaactcca atagcatgat tgatgctttc  1440
agtagaattt cctctggaac tggagacatt ttccagcaac atattcagct tgaaagtaca  1500
ggtgaaaatg tcaaacctca ccatcaattg aaaaacacag tgactgtgga taatactgtg  1560
```

```
ggcaacgaca ctatgtttct agttacgtgg caggccagtg gtcctcctga gattatatta    1620
tttgatcctg atggacgaaa atactacaca aataatttta tcaccaatct aactttcgg     1680
acagctagtc tttggattcc aggaacagct aagcctgggc actggactta caccctgaac    1740
aatacccatc attctctgca agccctgaaa gtgacagtga cctctcgcgc ctccaactca    1800
gctgtgcccc cagccactgt ggaagccttt gtggaaagag acagcctcca ttttcctcat    1860
cctgtgatga tttatgccaa tgtgaaacag ggattttatc ccattcttaa tgccactgtc    1920
actgccacag ttgagccaga gactggagat cctgttacgc tgagactcct tgatgatgga    1980
gcaggtgctg atgttataaa aaatgatgga atttactcga ggtattttt ctcctttgct     2040
gcaaatggta gatatagctt gaaagtgcat gtcaatcact ctcccagcat aagcacccca    2100
gcccactcta ttccagggag tcatgctatg tatgtaccag gttacacagc aaacggtaat    2160
attcagatga atgctccaag gaaatcagta ggcagaaatg aggaggagcg aaagtggggc    2220
tttagccgag tcagctcagg aggctccttt tcagtgctgg gagttccagc tggcccccac    2280
cctgatgtgt ttccaccatg caaaattatt gacctggaag ctgtaaaagt agaagaggaa    2340
ttgaccctat cttggacagc acctggagaa gactttgatc agggccaggc tacaagctat    2400
gaaataagaa tgagtaaaag tctacagaat atccaagatg actttaacaa tgctattta     2460
gtaaatacat caaagcgaaa tcctcagcaa gctggcatca gggagatatt tacgttctca    2520
ccccaaattt ccacgaatgg acctgaacat cagccaaatg gagaaacaca tgaaagccac    2580
agaatttatg ttgcaatacg agcaatggat aggaactcct tacagtctgc tgtatctaac    2640
attgccagg cgcctctgtt tattccccc aattctgatc ctgtacctgc cagagattat      2700
cttatattga aaggagtttt aacagcaatg ggtttgatag gaatcatttg ccttattata    2760
gttgtgacac atcatacttt aagcaggaaa aagagagcag acaagaaaga gaatggaaca    2820
aaattattat aa                                                        2832
```

Potein Sequence

Translation of human calcium-activated chloride channel, family member 2 (CLCA2) mRNA, ACCESSION NM_006536:

```
Met Thr Gln Arg Ser Ile Ala Gly Pro Ile Cys Asn Leu Lys Phe Val
1               5               10                  15

Thr Leu Leu Val Ala Leu Ser Ser Glu Leu Pro Phe Leu Gly Ala Gly
            20              25              30

Val Gln Leu Gln Asp Asn Gly Tyr Asn Gly Leu Leu Ile Ala Ile Asn
        35              40              45

Pro Gln Val Pro Glu Asn Gln Asn Leu Ile Ser Asn Ile Lys Glu Met
    50              55              60
```

APPENDIX

Ile Thr Glu Ala Ser Phe Tyr Leu Phe Asn Ala Thr Lys Arg Arg Val
65   70   75   80

Phe Phe Arg Asn Ile Lys Ile Leu Ile Pro Ala Thr Trp Lys Ala Asn
            85   90   95

Asn Asn Ser Lys Ile Lys Gln Glu Ser Tyr Glu Lys Ala Asn Val Ile
            100   105   110

Val Thr Asp Trp Tyr Gly Ala His Gly Asp Asp Pro Tyr Thr Leu Gln
        115   120   125

Tyr Arg Gly Cys Gly Lys Glu Gly Lys Tyr Ile His Phe Thr Pro Asn
        130   135   140

Phe Leu Leu Asn Asp Asn Leu Thr Ala Gly Tyr Gly Ser Arg Gly Arg
145   150   155   160

Val Phe Val His Glu Trp Ala His Leu Arg Trp Gly Val Phe Asp Glu
                165   170   175

Tyr Asn Asn Asp Lys Pro Phe Tyr Ile Asn Gly Gln Asn Gln Ile Lys
            180   185   190

Val Thr Arg Cys Ser Ser Asp Ile Thr Gly Ile Phe Val Cys Glu Lys
        195   200   205

Gly Pro Cys Pro Gln Glu Asn Cys Ile Ile Ser Lys Leu Phe Lys Glu
210   215   220

Gly Cys Thr Phe Ile Tyr Asn Ser Thr Gln Asn Ala Thr Ala Ser Ile
225   230   235   240

Met Phe Met Gln Ser Leu Ser Ser Val Val Glu Phe Cys Asn Ala Ser
            245   250   255

Thr His Asn Gln Glu Ala Pro Asn Leu Gln Asn Gln Met Cys Ser Leu
            260   265   270

Arg Ser Ala Trp Asp Val Ile Thr Asp Ser Ala Asp Phe His His Ser
        275   280   285

Phe Pro Met Asn Gly Thr Glu Leu Pro Pro Pro Thr Phe Ser Leu
        290   295   300

Val Gln Ala Gly Asp Lys Val Val Cys Leu Val Leu Asp Val Ser Ser
305   310   315   320

Lys Met Ala Glu Ala Asp Arg Leu Leu Gln Leu Gln Ala Ala Glu
            325   330   335

Phe Tyr Leu Met Gln Ile Val Glu Ile His Thr Phe Val Gly Ile Ala
            340   345   350

Ser Phe Asp Ser Lys Gly Glu Ile Arg Ala Gln Leu His Gln Ile Asn
        355   360   365

Ser Asn Asp Asp Arg Lys Leu Leu Val Ser Tyr Leu Pro Thr Thr Val
    370   375   380

Ser Ala Lys Thr Asp Ile Ser Ile Cys Ser Gly Leu Lys Lys Gly Phe
385   390   395   400

Glu Val Val Glu Lys Leu Asn Gly Lys Ala Tyr Gly Ser Val Met Ile
            405   410   415

Leu Val Thr Ser Gly Asp Asp Lys Leu Leu Gly Asn Cys Leu Pro Thr
            420   425   430

Val Leu Ser Ser Gly Ser Thr Ile His Ser Ile Ala Leu Gly Ser Ser
        435   440   445

Ala Ala Pro Asn Leu Glu Glu Leu Ser Arg Leu Thr Gly Gly Leu Lys
    450   455   460

Phe Phe Val Pro Asp Ile Ser Asn Ser Asn Ser Met Ile Asp Ala Phe
465   470   475   480

Ser Arg Ile Ser Ser Gly Thr Gly Asp Ile Phe Gln Gln His Ile Gln
            485   490   495

Leu Glu Ser Thr Gly Glu Asn Val Lys Pro His His Gln Leu Lys Asn
            500   505   510

Thr Val Thr Val Asp Asn Thr Val Gly Asn Asp Thr Met Phe Leu Val
            515   520   525

Thr Trp Gln Ala Ser Gly Pro Pro Glu Ile Ile Leu Phe Asp Pro Asp
    530   535   540

Gly Arg Lys Tyr Tyr Thr Asn Asn Phe Ile Thr Asn Leu Thr Phe Arg
545   550   555   560

Thr Ala Ser Leu Trp Ile Pro Gly Thr Ala Lys Pro Gly His Trp Thr
                565   570   575

Tyr Thr Leu Asn Asn Thr His His Ser Leu Gln Ala Leu Lys Val Thr
            580   585   590

Val Thr Ser Arg Ala Ser Asn Ser Ala Val Pro Pro Ala Thr Val Glu
        595   600   605

APPENDIX

```
Ala Phe Val Glu Arg Asp Ser Leu His Phe Pro His Pro Val Met Ile
    610              615              620
Tyr Ala Asn Val Lys Gln Gly Phe Tyr Pro Ile Leu Asn Ala Thr Val
625              630              635              640
Thr Ala Thr Val Glu Pro Glu Thr Gly Asp Pro Val Thr Leu Arg Leu
            645              650              655
Leu Asp Asp Gly Ala Gly Ala Asp Val Ile Lys Asn Asp Gly Ile Tyr
            660              665              670
Ser Arg Tyr Phe Phe Ser Phe Ala Ala Asn Gly Arg Tyr Ser Leu Lys
        675              680              685
Val His Val Asn His Ser Pro Ser Ile Ser Thr Pro Ala His Ser Ile
    690              695              700
Pro Gly Ser His Ala Met Tyr Val Pro Gly Tyr Thr Ala Asn Gly Asn
705              710              715              720
Ile Gln Met Asn Ala Pro Arg Lys Ser Val Gly Arg Asn Glu Glu Glu
                725              730              735
Arg Lys Trp Gly Phe Ser Arg Val Ser Ser Gly Ser Phe Ser Val
            740              745              750
Leu Gly Val Pro Ala Gly Pro His Pro Asp Val Phe Pro Pro Cys Lys
        755              760              765
Ile Ile Asp Leu Glu Ala Val Lys Val Glu Glu Leu Thr Leu Ser
    770              775              780
Trp Thr Ala Pro Gly Glu Asp Phe Asp Gln Gly Gln Ala Thr Ser Tyr
785              790              795              800
Glu Ile Arg Met Ser Lys Ser Leu Gln Asn Ile Gln Asp Asp Phe Asn
            805              810              815
Asn Ala Ile Leu Val Asn Thr Ser Lys Arg Asn Pro Gln Gln Ala Gly
            820              825              830
Ile Arg Glu Ile Phe Thr Phe Ser Pro Gln Ile Ser Thr Asn Gly Pro
        835              840              845
Glu His Gln Pro Asn Gly Glu Thr His Glu Ser His Arg Ile Tyr Val
    850              855              860
Ala Ile Arg Ala Met Asp Arg Asn Ser Leu Gln Ser Ala Val Ser Asn
865              870              875              880
```

APPENDIX

Ile Ala Gln Ala Pro Leu Phe Ile Pro Pro Asn Ser Asp Pro Val Pro
            885   890   895

Ala Arg Asp Tyr Leu Ile Leu Lys Gly Val Leu Thr Ala Met Gly Leu
            900   905   910

Ile Gly Ile Ile Cys Leu Ile Ile Val Val Thr His His Thr Leu Ser
            915   920   925

Arg Lys Lys Arg Ala Asp Lys Lys Glu Asn Gly Thr Lys Leu Leu
            930   935   940

Abbreviations

°C	degree Celsius
µg	microgram
µl	microliter
µmol	micromole
14-3-3 sigma/ SDN	stratifin
2D	2 dimensional
3D	3 dimensional
4E-BP1	4E binding protein 1
AA	amino acid
AC	adenocarcinoma
Acc.	accession number
AFF4 (MCEF)	AF4/MR2 family member 4
Ala	Alanine
All	acute lymphocytic leukaemia
amp	Ampicillin
ASPM	Asp (abnormal spindle)-like, microcephaly associated (drosophila)
ATP	adenosine 5'-triphosphate
bp	base pair
BSA	bovine serum albumin
C10orf118	chromosome 10 open reading frame 118
Ca	carcinoma
Ca^{2+}	calcium ion
CAM's	cell-cell adhesion molecules
cap	methyl-7-G(5')pppN - structure
CD8	cluster of differentiation
CDIPT	CDP-diacylglycerol-inositol 3-phosphatidyltransferase (phosphatidyinositol synthase)
CDK	cyclin-dependent kinase
cDNA	complementary DNA
cfu	colony forming unit
CHD6	chromodomain helicase DNA binding protein 6
chr1	Chromosome 1
CLCA	calcium-activated chloride channel
CMTX	Charcot-Marie-Tooth disease
CMV	cytomegalo virus
CN	copy number
ColonMucinCa	mucinous colon carcinoma
COPD	obstructive pulmonary disease
cpm	counts per minute
CTCL	cutaneous T cell lymphoma
Cys	Cysteine
DEPC	diethyl pyrocarbonate
DEPC-H_2O	di-ethyl-propyl carbonate-treated H_2O
DNA	deoxyribonucleic acid
dNTP	deoxyribonucleotides
Doxy	Doxycycline
DSC3	desmocollin 3

dsDNA	double strand DNA
DSG3	desmoglein 3
DTT	dithiothreitol
DTX2	deltex homolog 2 (drosophila)
E. coli	Escherichia coli
ECM	extracellular matrix
EDTA	ethylenediaminetetraacetic acid
EGF	epidermal growth factor
EGFR	epidermal growth factor receptor
eIF	eukaryotic initiation factor
EMCV	encephalomyocraditis virus
EMT	epithilial mesenchymal transition
EP300	E1A binding protein p300
ERBB2	erythroblastic leukemia viral oncogene homolog 2
ERK	extracellular signal-regulated kinase
EtBr	ethidiumbromide
FGF	fibroblast growth factor
g	gram
GAPDH	glyceraldehyde 3-phosphate dehydrogenase
GDP	guanosine 5'-diphosphate
GlioblastMultiforme	glioblastoma multiforme
GPCR	G protein-coupled receptor
GTP	guanosine 5'-triphosphate
HA	hemagglutinin
HAV	hepatitis A virus
HCC	hepatocellular carcinoma
HCS	high-content screening
Her2	epidermal growth factor receptor 2
His	Histidine
HNF4A	hepatocyte nuclear factor 4 alpha
HNSCC	head and neck squamous cell carcinoma
HRV2	human rhinovirus type 2
IC_{50}	half maximal inhibitory concentration
IGF-I R	insulin-like growth factor 1 receptor
IHC	immunohistochemistry
IL10	Interleukin 10
IL2	Interleukin 2
InfDuc	infiltrating ductal
InfLob	infiltrating lobular
IPA	Ingenuity Pathways Analysis
IQR	interquartile range
JNK	N-terminal kinase
K^+	potassium ion
KAc	potassium acetate
kD	kilo Dalton
KPNA2	karyopherin alpha 2
KRT5	keratin 5
l	liter
LAD1/CD18	ladinin 1
LB	Lauria Bertani
LCC	large cell carcinomas

APPENDIX

LCM	laser-capture-microdissected
LN	lymphnode
LNSCCMet	lymph node metastases of SCCs
LTBP1	latent transforming growth factor beta binding protein 1
Lys	Lysine
M	molar
MACF1	microtuble-actin crosslinking factor 1
Malig	malignant
MAPK	mitogen-activated protein kinase
Mel	melanoma
MEM	minimal essential medium
Met	metastasis
min	minute
ml	milliliter
MLL2	myeloid/lymphoid or mixed-lineage leukemia 2
mM	millimolare
MM	multiple myeloma
MOPS	4-morpholinepropanesulfonic acid
MRE11	meiotic recombination 11, S. cerevisiae, homolog of a MRE11A
mRNA	messenger RNA
myc	refers to the myc gene
Myc	refers to the Myc protein
NaAc	sodium acetate
NBS1	p95 protein of the MRE11/RAD50 complex
NEAA	non essential aminoacids
NED	no evidence of disease
ng	nanogram
NH_4Ac	ammonium acetate
NPM	nucleophosmin
NSCLC	non-small cell lung cancer
nt	nucleotide
OD	optical density
ORF	open reading frame
P4HA2	prolyl-4 hydroxylase-2
p53	protein 53
PapillFollic	papillary and follicular
PARP	poly ADP-ribose polymerase
PBS	phosphate-buffered saline
PCR	polymerase chain reaction
PDGF	platelet-derived growth factor
PDK1	phosphoinositide dependent kinase 1
PERP	TP53 apoptosis effector
Phe	Phenylalanine
PI3K	phosphoinositide-3-kinase
PKP	plakophilin
pmol	picomole
POS	positive
PV	verified by pathologist
RAB38	member RAS oncogene family
RAD50	RAD50 homolog (S. cerevisiae)
Rb	retinoblastoma

APPENDIX

RCC	renal cell carcinoma
RE	restriction enzyme
RNA	ribonucleic acid
RNAse	ribonuclease
rpm	rotations per minute
RT	room temperature
RTK	receptor tyrosine kinase
RT-qPCR	quantative real-time PCR
RUTBC3	RUN and TBC1 domain containing 3
SATB1	special AT-rich sequence binding protein 1
SCC	squamous cell carcinoma
SCLC	small cell lang cancer
SDS	sodium dodecyl sulfate
sec	second
SFI1	spindle assembly associated (yeast)
siCTRL	non-targeting siRNA CONTROL#1 of Dharmacon
siRNA	small interfering RNA (also short interfering RNA or silencing RNA)
SMAD3	mothers against decapentaplegic homolog 3
SNP	single nucleotide polymorphism
Squam	squamous
ssDNA	single strand cDNA
STAT	signal transducer and activator of transcription
T reg	regulatory T cells
TCC	transitional cell carcinoma
Tet	Tetracycline
TGF	tumour growth factor
Tiss	tissue
Tm	melting temperature
TNF	tumour necrosis factor
TP53	tumour protein p53
TP53BP1	tumour protein p53 binding protein 1
TRIM29	tripartite motif-containing 29
TSC1	tuberous sclerosis complex 1
Tyr	Tyrosine
U	Unit
UNC84A	unc-84 homolog A (C. elegans)
UV	ultraviolet
VEGF	vascular endothelial growth factor
VP2	adeno-associated virus 2
vWA	von Willebrand factor type A
WHO	World Health Organization
wt	wild type
Zn^{2+}	zinc ion

REFERENCES

Reference List

(1) Amatschek S, Koenig U, Auer H, Steinlein P, Pacher M, Gruenfelder A, Dekan G, Vogl S, Kubista E, Heider KH, Stratowa C, Schreiber M, Sommergruber W. Tissue-wide expression profiling using cDNA subtraction and microarrays to identify tumor-specific genes. Cancer Res 2004; 64(3):844-856.

(2) Konopitzky R, Konig U, Meyer RG, Sommergruber W, Wolfel T, Schweighoffer T. Identification of HLA-A*0201-restricted T cell epitopes derived from the novel overexpressed tumor antigen calcium-activated chloride channel 2. J Immunol 2002; 169(1):540-547.

(3) Fuller CM, Ji HL, Tousson A, Elble RC, Pauli BU, Benos DJ. Ca(2+)-activated Cl(-) channels: a newly emerging anion transport family. Pflugers Arch 2001; 443 Suppl 1:S107-S110.

(4) Pauli BU, Abdel-Ghany M, Cheng HC, Gruber AD, Archibald HA, Elble RC. Molecular characteristics and functional diversity of CLCA family members. Clin Exp Pharmacol Physiol 2000; 27(11):901-905.

(5) Jemal A, Siegel R, Ward E, Hao Y, Xu J, Murray T, Thun MJ. Cancer statistics, 2008. CA Cancer J Clin 2008; 58(2):71-96.

(6) Ames BN. Identifying environmental chemicals causing mutations and cancer. Science 1979; 204(4393):587-593.

(7) Schmutte C, Fishel R. Genomic instability: first step to carcinogenesis. Anticancer Res 1999; 19(6A):4665-4696.

(8) Anderson J, Gordon A, McManus A, Shipley J, Pritchard-Jones K. Disruption of imprinted genes at chromosome region 11p15.5 in paediatric rhabdomyosarcoma. Neoplasia 1999; 1(4):340-348.

(9) Herman JG, Baylin SB. Promoter-region hypermethylation and gene silencing in human cancer. Curr Top Microbiol Immunol 2000; 249:35-54.

(10) Sigurdsson S, Bodvarsdottir SK, Anamthawat-Jonsson K, Steinarsdottir M, Jonasson JG, Ogmundsdottir HM, Eyfjord JE. p53 abnormality and chromosomal instability in the same breast tumor cells. Cancer Genet Cytogenet 2000; 121(2):150-155.

(11) Brueckl WM, Jung A, Wein A, Brabletz T, Guenther K, Nusko G, Hahn EG. Microsatellite instability in colorectal adenomas: relevance and clinical importance. Int J Colorectal Dis 2000; 15(4):189-196.

REFERENCES

(12) Nowell PC. The clonal evolution of tumor cell populations. Science 1976; 194(4260):23-28.

(13) Weinberg RA. Oncogenes, antioncogenes, and the molecular bases of multistep carcinogenesis. Cancer Res 1989; 49(14):3713-3721.

(14) Vogelstein B, Kinzler KW. The multistep nature of cancer. Trends Genet 1993; 9(4):138-141.

(15) Weinstein IB. Cancer. Addiction to oncogenes--the Achilles heal of cancer. Science 2002; 297(5578):63-64.

(16) Hanahan D, Weinberg RA. The hallmarks of cancer. Cell 2000; 100(1):57-70.

(17) Fedi P, Tronick SR, Aaronson SA. Growth factors in cancer medicine eds 1997.

(18) Datto MB, Hu PP, Kowalik TF, Yingling J, Wang XF. The viral oncoprotein E1A blocks transforming growth factor beta-mediated induction of p21/WAF1/Cip1 and p15/INK4B. Mol Cell Biol 1997; 17(4):2030-2037.

(19) Hannon GJ, Beach D. p15INK4B is a potential effector of TGF-beta-induced cell cycle arrest. Nature 1994; 371(6494):257-261.

(20) Hahne M, Rimoldi D, Schroter M, Romero P, Schreier M, French LE, Schneider P, Bornand T, Fontana A, Lienard D, Cerottini J, Tschopp J. Melanoma cell expression of Fas(Apo-1/CD95) ligand: implications for tumor immune escape. Science 1996; 274(5291):1363-1366.

(21) O'Connell J, O'Sullivan GC, Collins JK, Shanahan F. The Fas counterattack: Fas-mediated T cell killing by colon cancer cells expressing Fas ligand. J Exp Med 1996; 184(3):1075-1082.

(22) Shay JW, Bacchetti S. A survey of telomerase activity in human cancer. Eur J Cancer 1997; 33(5):787-791.

(23) Hanahan D, Folkman J. Patterns and emerging mechanisms of the angiogenic switch during tumorigenesis. Cell 1996; 86(3):353-364.

(24) Aplin AE, Howe A, Alahari SK, Juliano RL. Signal transduction and signal modulation by cell adhesion receptors: the role of integrins, cadherins, immunoglobulin-cell adhesion molecules, and selectins. Pharmacol Rev 1998; 50(2):197-263.

(25) Christofori G, Semb H. The role of the cell-adhesion molecule E-cadherin as a tumour-suppressor gene. Trends Biochem Sci 1999; 24(2):73-76.

(26) Georgolios A, Batistatou A, Manolopoulos L, Charalabopoulos K. Role and expression patterns of E-cadherin in head and neck squamous cell carcinoma (HNSCC). J Exp Clin Cancer Res 2006; 25(1):5-14.

REFERENCES

(27) Eriksen JG, Steiniche T, Sogaard H, Overgaard J. Expression of integrins and E-cadherin in squamous cell carcinomas of the head and neck. APMIS 2004; 112(9):560-568.

(28) Weinstein IB, Joe A. Oncogene addiction. Cancer Res 2008; 68(9):3077-3080.

(29) Hubner A, Jaeschke A, Davis RJ. Oncogene addiction: role of signal attenuation. Dev Cell 2006; 11(6):752-754.

(30) Paez JG, Janne PA, Lee JC, Tracy S, Greulich H, Gabriel S, Herman P, Kaye FJ, Lindeman N, Boggon TJ, Naoki K, Sasaki H, Fujii Y, Eck MJ, Sellers WR, Johnson BE, Meyerson M. EGFR mutations in lung cancer: correlation with clinical response to gefitinib therapy. Science 2004; 304(5676):1497-1500.

(31) Zhang Z, Li M, Rayburn ER, Hill DL, Zhang R, Wang H. Oncogenes as novel targets for cancer therapy (part I): growth factors and protein tyrosine kinases. Am J Pharmacogenomics 2005; 5(3):173-190.

(32) Nuber UA, Schafer S, Stehr S, Rackwitz HR, Franke WW. Patterns of desmocollin synthesis in human epithelia: immunolocalization of desmocollins 1 and 3 in special epithelia and in cultured cells. Eur J Cell Biol 1996; 71(1):1-13.

(33) Bonifas JM, Bare JW, Lynch ED, Lebo RV, Epstein EH, Jr. Regional assignment of the human keratin 5 (KRT5) gene to chromosome 12q near D12S14 by PCR analysis of somatic cell hybrids and multicolor in situ hybridization. Genomics 1992; 13(2):452-454.

(34) Mahoney MG, Simpson A, Aho S, Uitto J, Pulkkinen L. Interspecies conservation and differential expression of mouse desmoglein gene family. Exp Dermatol 2002; 11(2):115-125.

(35) Heid HW, Schmidt A, Zimbelmann R, Schafer S, Winter-Simanowski S, Stumpp S, Keith M, Figge U, Schnolzer M, Franke WW. Cell type-specific desmosomal plaque proteins of the plakoglobin family: plakophilin 1 (band 6 protein). Differentiation 1994; 58(2):113-131.

(36) Schmidt A, Langbein L, Pratzel S, Rode M, Rackwitz HR, Franke WW. Plakophilin 3--a novel cell-type-specific desmosomal plaque protein. Differentiation 1999; 64(5):291-306.

(37) Ihrie RA, Marques MR, Nguyen BT, Horner JS, Papazoglu C, Bronson RT, Mills AA, Attardi LD. Perp is a p63-regulated gene essential for epithelial integrity. Cell 2005; 120(6):843-856.

REFERENCES

(38) Sanderson RD, Bernfield M. Molecular polymorphism of a cell surface proteoglycan: distinct structures on simple and stratified epithelia. Proc Natl Acad Sci U S A 1988; 85(24):9562-9566.

(39) Johnson TM, Rowe DE, Nelson BR, Swanson NA. Squamous cell carcinoma of the skin (excluding lip and oral mucosa). J Am Acad Dermatol 1992; 26(3 Pt 2):467-484.

(40) Christenson LJ, Borrowman TA, Vachon CM, Tollefson MM, Otley CC, Weaver AL, Roenigk RK. Incidence of basal cell and squamous cell carcinomas in a population younger than 40 years. JAMA 2005; 294(6):681-690.

(41) Tang CH, Chuang JY, Fong YC, Maa MC, Way TD, Hung CH. Bone-derived SDF-1 stimulates IL-6 release via CXCR4, ERK and NF-kappaB pathways and promotes osteoclastogenesis in human oral cancer cells. Carcinogenesis 2008; 29(8):1483-1492.

(42) Rapidis AD, Vermorken JB, Bourhis J. Targeted therapies in head and neck cancer: past, present and future. Rev Recent Clin Trials 2008; 3(3):156-166.

(43) Cunningham SA, Awayda MS, Bubien JK, Ismailov II, Arrate MP, Berdiev BK, Benos DJ, Fuller CM. Cloning of an epithelial chloride channel from bovine trachea. J Biol Chem 1995; 270(52):31016-31026.

(44) Ran S, Benos DJ. Isolation and functional reconstitution of a 38-kDa chloride channel protein from bovine tracheal membranes. J Biol Chem 1991; 266(8):4782-4788.

(45) Zhu DZ, Cheng CF, Pauli BU. Mediation of lung metastasis of murine melanomas by a lung-specific endothelial cell adhesion molecule. Proc Natl Acad Sci U S A 1991; 88(21):9568-9572.

(46) Elble RC, Widom J, Gruber AD, Abdel-Ghany M, Levine R, Goodwin A, Cheng HC, Pauli BU. Cloning and characterization of lung-endothelial cell adhesion molecule-1 suggest it is an endothelial chloride channel. J Biol Chem 1997; 272(44):27853-27861.

(47) Gandhi R, Elble RC, Gruber AD, Schreur KD, Ji HL, Fuller CM, Pauli BU. Molecular and functional characterization of a calcium-sensitive chloride channel from mouse lung. J Biol Chem 1998; 273(48):32096-32101.

(48) Romio L, Musante L, Cinti R, Seri M, Moran O, Zegarra-Moran O, Galietta LJ. Characterization of a murine gene homologous to the bovine CaCC chloride channel. Gene 1999; 228(1-2):181-188.

(49) Lee D, Ha S, Kho Y, Kim J, Cho K, Baik M, Choi Y. Induction of mouse Ca(2+)-sensitive chloride channel 2 gene during involution of mammary gland. Biochem Biophys Res Commun 1999; 264(3):933-937.

REFERENCES

(50) Komiya T, Tanigawa Y, Hirohashi S. Cloning and identification of the gene gob-5, which is expressed in intestinal goblet cells in mice. Biochem Biophys Res Commun 1999; 255(2):347-351.

(51) Elble RC, Ji G, Nehrke K, DeBiasio J, Kingsley PD, Kotlikoff MI, Pauli BU. Molecular and functional characterization of a murine calcium-activated chloride channel expressed in smooth muscle. J Biol Chem 2002; 277(21):18586-18591.

(52) Evans SR, Thoreson WB, Beck CL. Molecular and functional analyses of two new calcium-activated chloride channel family members from mouse eye and intestine. J Biol Chem 2004; 279(40):41792-41800.

(53) Gruber AD, Elble RC, Ji HL, Schreur KD, Fuller CM, Pauli BU. Genomic cloning, molecular characterization, and functional analysis of human CLCA1, the first human member of the family of Ca2+-activated Cl- channel proteins. Genomics 1998; 54(2):200-214.

(54) Agnel M, Vermat T, Culouscou JM. Identification of three novel members of the calcium-dependent chloride channel (CaCC) family predominantly expressed in the digestive tract and trachea. FEBS Lett 1999; 455(3):295-301.

(55) Gruber AD, Schreur KD, Ji HL, Fuller CM, Pauli BU. Molecular cloning and transmembrane structure of hCLCA2 from human lung, trachea, and mammary gland. Am J Physiol 1999; 276(6 Pt 1):C1261-C1270.

(56) Itoh R, Kawamoto S, Miyamoto Y, Kinoshita S, Okubo K. Isolation and characterization of a Ca(2+)-activated chloride channel from human corneal epithelium. Curr Eye Res 2000; 21(6):918-925.

(57) Gaspar KJ, Racette KJ, Gordon JR, Loewen ME, Forsyth GW. Cloning a chloride conductance mediator from the apical membrane of porcine ileal enterocytes. Physiol Genomics 2000; 3(2):101-111.

(58) Yamazaki J, Okamura K, Ishibashi K, Kitamura K. Characterization of CLCA protein expressed in ductal cells of rat salivary glands. Biochim Biophys Acta 2005; 1715(2):132-144.

(59) Jeong SM, Park HK, Yoon IS, Lee JH, Kim JH, Jang CG, Lee CJ, Nah SY. Cloning and expression of Ca2+-activated chloride channel from rat brain. Biochem Biophys Res Commun 2005; 334(2):569-576.

(60) Yoon IS, Jeong SM, Lee SN, Lee JH, Kim JH, Pyo MK, Lee JH, Lee BH, Choi SH, Rhim H, Choe H, Nah SY. Cloning and heterologous expression of a Ca2+-activated chloride channel isoform from rat brain. Biol Pharm Bull 2006; 29(11):2168-2173.

REFERENCES

(61) Anton F, Leverkoehne I, Mundhenk L, Thoreson WB, Gruber AD. Overexpression of eCLCA1 in small airways of horses with recurrent airway obstruction. J Histochem Cytochem 2005; 53(8):1011-1021.

(62) Loewen ME, Smith NK, Hamilton DL, Grahn BH, Forsyth GW. CLCA protein and chloride transport in canine retinal pigment epithelium. Am J Physiol Cell Physiol 2003; 285(5):C1314-C1321.

(63) Gruber AD, Pauli BU. Clustering of the human CLCA gene family on the short arm of chromosome 1 (1p22-31). Genome 1999; 42(5):1030-1032.

(64) Gruber AD, Pauli BU. Molecular cloning and biochemical characterization of a truncated, secreted member of the human family of Ca2+-activated Cl-channels. Biochim Biophys Acta 1999; 1444(3):418-423.

(65) Pawlowski K, Lepisto M, Meinander N, Sivars U, Varga M, Wieslander E. Novel conserved hydrolase domain in the CLCA family of alleged calcium-activated chloride channels. Proteins 2006; 63(3):424-439.

(66) Elble RC, Pauli BU. Tumor suppression by a proapoptotic calcium-activated chloride channel in mammary epithelium. J Biol Chem 2001; 276(44):40510-40517.

(67) Elble RC, Walia V, Cheng HC, Connon CJ, Mundhenk L, Gruber AD, Pauli BU. The putative chloride channel hCLCA2 has a single C-terminal transmembrane segment. J Biol Chem 2006; 281(40):29448-29454.

(68) Gruber AD, Pauli BU. Tumorigenicity of human breast cancer is associated with loss of the Ca2+-activated chloride channel CLCA2. Cancer Res 1999; 59(21):5488-5491.

(69) Abdel-Ghany M, Cheng HC, Elble RC, Lin H, DiBiasio J, Pauli BU. The interacting binding domains of the beta(4) integrin and calcium-activated chloride channels (CLCAs) in metastasis. J Biol Chem 2003; 278(49):49406-49416.

(70) Abdel-Ghany M, Cheng HC, Elble RC, Pauli BU. The breast cancer beta 4 integrin and endothelial human CLCA2 mediate lung metastasis. J Biol Chem 2001; 276(27):25438-25446.

(71) Kumanovics A, Lindahl KF. G7c in the lung tumor susceptibility (Lts) region of the Mhc class III region encodes a von Willebrand factor type A domain protein. Immunogenetics 2001; 53(1):64-68.

(72) Abdel-Ghany M, Cheng HC, Elble RC, Pauli BU. Focal adhesion kinase activated by beta(4) integrin ligation to mCLCA1 mediates early metastatic growth. J Biol Chem 2002; 277(37):34391-34400.

REFERENCES

(73) Goetz DJ, el Sabban ME, Hammer DA, Pauli BU. Lu-ECAM-1-mediated adhesion of melanoma cells to endothelium under conditions of flow. Int J Cancer 1996; 65(2):192-199.

(74) Riker AI, Enkemann SA, Fodstad O, Liu S, Ren S, Morris C, Xi Y, Howell P, Metge B, Samant RS, Shevde LA, Li W, Eschrich S, Daud A, Ju J, Matta J. The gene expression profiles of primary and metastatic melanoma yields a transition point of tumor progression and metastasis. BMC Med Genomics 2008; 1:13.

(75) Shen-Ong GL, Feng Y, Troyer DA. Expression profiling identifies a novel alpha-methylacyl-CoA racemase exon with fumarate hydratase homology. Cancer Res 2003; 63(12):3296-3301.

(76) Tovey SM, Brown S, Doughty JC, Mallon EA, Cooke TG, Edwards J. Poor survival outcomes in HER2-positive breast cancer patients with low-grade, node-negative tumours. Br J Cancer 2009; 100(5):680-683.

(77) Liu S, Chia SK, Mehl E, Leung S, Rajput A, Cheang MC, Nielsen TO. Progesterone receptor is a significant factor associated with clinical outcomes and effect of adjuvant tamoxifen therapy in breast cancer patients. Breast Cancer Res Treat 2009.

(78) Hupe P, Stransky N, Thiery JP, Radvanyi F, Barillot E. Analysis of array CGH data: from signal ratio to gain and loss of DNA regions. Bioinformatics 2004; 20(18):3413-3422.

(79) Glatt S, Halbauer D, Heindl S, Wernitznig A, Kozina D, Su KC, Puri C, Garin-Chesa P, Sommergruber W. hGPR87 contributes to viability of human tumor cells. Int J Cancer 2008; 122(9):2008-2016.

(80) Chou PY, Fasman GD. Empirical predictions of protein conformation. Annu Rev Biochem 1978; 47:251-276.

(81) Chou PY, Fasman GD. Prediction of the secondary structure of proteins from their amino acid sequence. Adv Enzymol Relat Areas Mol Biol 1978; 47:45-148.

(82) Garnier J, Osguthorpe DJ, Robson B. Analysis of the accuracy and implications of simple methods for predicting the secondary structure of globular proteins. J Mol Biol 1978; 120(1):97-120.

(83) Kyte J, Doolittle RF. A simple method for displaying the hydropathic character of a protein. J Mol Biol 1982; 157(1):105-132.

(84) von Heijne G. On the hydrophobic nature of signal sequences. Eur J Biochem 1981; 116(2):419-422.

REFERENCES

(85) Engelman DM, Steitz TA, Goldman A. Identifying nonpolar transbilayer helices in amino acid sequences of membrane proteins. Annu Rev Biophys Biophys Chem 1986; 15:321-353.

(86) Karplus PA, Schulz GE. Prediction of chain flexibility in proteins: a tool for the selection of peptide antigens. Naturwissenschaften 1985; 72:212-213.

(87) Hopp TP, Woods KR. Prediction of protein antigenic determinants from amino acid sequences. Proc Natl Acad Sci U S A 1981; 78(6):3824-3828.

(88) Parker JM, Guo D, Hodges RS. New hydrophilicity scale derived from high-performance liquid chromatography peptide retention data: correlation of predicted surface residues with antigenicity and X-ray-derived accessible sites. Biochemistry 1986; 25(19):5425-5432.

(89) Thornton JM, Edwards MS, Taylor WR, Barlow DJ. Location of 'continuous' antigenic determinants in the protruding regions of proteins. EMBO J 1986; 5(2):409-413.

(90) Welling GW, Weijer WJ, van der ZR, Welling-Wester S. Prediction of sequential antigenic regions in proteins. FEBS Lett 1985; 188(2):215-218.

(91) Alix AJ. Predictive estimation of protein linear epitopes by using the program PEOPLE. Vaccine 1999; 18(3-4):311-314.

(92) Jameson BA, Wolf H. The antigenic index: a novel algorithm for predicting antigenic determinants. Comput Appl Biosci 1988; 4(1):181-186.

(93) Bairoch A, Bucher P, Hofmann K. The PROSITE database, its status in 1997. Nucleic Acids Res 1997; 25(1):217-221.

(94) Skern T, Neubauer C, Frasel L, Grundler P, Sommergruber W, Zorn M, Kuechler E, Blaas D. A neutralizing epitope on human rhinovirus type 2 includes amino acid residues between 153 and 164 of virus capsid protein VP2. J Gen Virol 1987; 68 (Pt 2):315-323.

(95) Devi L. Consensus sequence for processing of peptide precursors at monobasic sites. FEBS Lett 1991; 280(2):189-194.

(96) Salisbury JL. Centrosomes: Sfi1p and centrin unravel a structural riddle. Curr Biol 2004; 14(1):R27-R29.

(97) Naruke Y, Nakashima M, Suzuki K, Matsuu-Matsuyama M, Shichijo K, Kondo H, Sekine I. Alteration of p53-binding protein 1 expression during skin carcinogenesis: association with genomic instability. Cancer Sci 2008; 99(5):946-951.

(98) Baba Y, Tsukuda M, Mochimatsu I, Furukawa S, Kagata H, Nagashima Y, Koshika S, Imoto M, Kato Y. Cytostatic effect of inostamycin, an inhibitor of

cytidine 5'-diphosphate 1,2-diacyl-sn-glycerol (CDP-DG): inositol transferase, on oral squamous cell carcinoma cell lines. Cell Biol Int 2001; 25(7):613-620.

(99) Rohatgi N, Matta A, Kaur J, Srivastava A, Ralhan R. Novel molecular targets of smokeless tobacco (khaini) in cell culture from oral hyperplasia. Toxicology 2006; 224(1-2):1-13.

(100) Issaeva I, Zonis Y, Rozovskaia T, Orlovsky K, Croce CM, Nakamura T, Mazo A, Eisenbach L, Canaani E. Knockdown of ALR (MLL2) reveals ALR target genes and leads to alterations in cell adhesion and growth. Mol Cell Biol 2007; 27(5):1889-1903.

(101) Bach C, Mueller D, Buhl S, Garcia-Cuellar MP, Slany RK. Alterations of the CxxC domain preclude oncogenic activation of mixed-lineage leukemia 2. Oncogene 2009; 28(6):815-823.

(102) Robert I, Aussems M, Keutgens A, Zhang X, Hennuy B, Viatour P, Vanstraelen G, Merville MP, Chapelle JP, de Leval L, Lambert F, Dejardin E, Gothot A, Chariot A. Matrix Metalloproteinase-9 gene induction by a truncated oncogenic NF-kappaB2 protein involves the recruitment of MLL1 and MLL2 H3K4 histone methyltransferase complexes. Oncogene 2009.

(103) Estable MC, Naghavi MH, Kato H, Xiao H, Qin J, Vahlne A, Roeder RG. MCEF, the newest member of the AF4 family of transcription factors involved in leukemia, is a positive transcription elongation factor-b-associated protein. J Biomed Sci 2002; 9(3):234-245.

(104) Hasan S, Guttinger S, Muhlhausser P, Anderegg F, Burgler S, Kutay U. Nuclear envelope localization of human UNC84A does not require nuclear lamins. FEBS Lett 2006; 580(5):1263-1268.

(105) Wilkin M, Tongngok P, Gensch N, Clemence S, Motoki M, Yamada K, Hori K, Taniguchi-Kanai M, Franklin E, Matsuno K, Baron M. Drosophila HOPS and AP-3 complex genes are required for a Deltex-regulated activation of notch in the endosomal trafficking pathway. Dev Cell 2008; 15(5):762-772.

(106) Jennings MD, Blankley RT, Baron M, Golovanov AP, Avis JM. Specificity and autoregulation of Notch binding by tandem WW domains in suppressor of Deltex. J Biol Chem 2007; 282(39):29032-29042.

(107) Han HJ, Russo J, Kohwi Y, Kohwi-Shigematsu T. SATB1 reprogrammes gene expression to promote breast tumour growth and metastasis. Nature 2008; 452(7184):187-193.

(108) Ghosh AK, Varga J. The transcriptional coactivator and acetyltransferase p300 in fibroblast biology and fibrosis. J Cell Physiol 2007; 213(3):663-671.

(109) Higashi T, Sasagawa T, Inoue M, Oka R, Shuangying L, Saijoh K. Overexpression of latent transforming growth factor-beta 1 (TGF-beta 1)

REFERENCES

binding protein 1 (LTBP-1) in association with TGF-beta 1 in ovarian carcinoma. Jpn J Cancer Res 2001; 92(5):506-515.

(110) Chakraborty S, Mohiyuddin SM, Gopinath KS, Kumar A. Involvement of TSC genes and differential expression of other members of the mTOR signaling pathway in oral squamous cell carcinoma. BMC Cancer 2008; 8:163.

(111) Maiuri MC, Tasdemir E, Criollo A, Morselli E, Vicencio JM, Carnuccio R, Kroemer G. Control of autophagy by oncogenes and tumor suppressor genes. Cell Death Differ 2009; 16(1):87-93.

(112) Chen HJ, Lin CM, Lin CS, Perez-Olle R, Leung CL, Liem RK. The role of microtubule actin cross-linking factor 1 (MACF1) in the Wnt signaling pathway. Genes Dev 2006; 20(14):1933-1945.

(113) Lutz T, Stoger R, Nieto A. CHD6 is a DNA-dependent ATPase and localizes at nuclear sites of mRNA synthesis. FEBS Lett 2006; 580(25):5851-5857.

(114) Katoh M, Katoh M. Characterization of RUSC1 and RUSC2 genes in silico. Oncol Rep 2004; 12(4):933-938.

(115) Callebaut I, de Gunzburg J, Goud B, Mornon JP. RUN domains: a new family of domains involved in Ras-like GTPase signaling. Trends Biochem Sci 2001; 26(2):79-83.

(116) Mackay A, Jones C, Dexter T, Silva RL, Bulmer K, Jones A, Simpson P, Harris RA, Jat PS, Neville AM, Reis LF, Lakhani SR, O'Hare MJ. cDNA microarray analysis of genes associated with ERBB2 (HER2/neu) overexpression in human mammary luminal epithelial cells. Oncogene 2003; 22(17):2680-2688.

(117) Jarzab B, Wiench M, Fujarewicz K, Simek K, Jarzab M, Oczko-Wojciechowska M, Wloch J, Czarniecka A, Chmielik E, Lange D, Pawlaczek A, Szpak S, Gubala E, Swierniak A. Gene expression profile of papillary thyroid cancer: sources of variability and diagnostic implications. Cancer Res 2005; 65(4):1587-1597.

(118) Lin SY, Pan HW, Liu SH, Jeng YM, Hu FC, Peng SY, Lai PL, Hsu HC. ASPM is a novel marker for vascular invasion, early recurrence, and poor prognosis of hepatocellular carcinoma. Clin Cancer Res 2008; 14(15):4814-4820.

(119) Clark RA, Huang SJ, Murphy GF, Mollet IG, Hijnen D, Muthukuru M, Schanbacher CF, Edwards V, Miller DM, Kim JE, Lambert J, Kupper TS. Human squamous cell carcinomas evade the immune response by down-regulation of vascular E-selectin and recruitment of regulatory T cells. J Exp Med 2008; 205(10):2221-2234.

(120) Di Fiore PP. Playing both sides: nucleophosmin between tumor suppression and oncogenesis. J Cell Biol 2008; 182(1):7-9.

REFERENCES

(121) Mishra BK, Purvish MP. Targeted Therapy in Oncology. MJAFI 2006; 62(2):169-173.

(122) Nielsen DL, Andersson M, Kamby C. HER2-targeted therapy in breast cancer. Monoclonal antibodies and tyrosine kinase inhibitors. Cancer Treat Rev 2008.

(123) Traxler P. Tyrosine kinases as targets in cancer therapy - successes and failures. Expert Opin Ther Targets 2003; 7(2):215-234.

(124) Li D, Ambrogio L, Shimamura T, Kubo S, Takahashi M, Chirieac LR, Padera RF, Shapiro GI, Baum A, Himmelsbach F, Rettig WJ, Meyerson M, Solca F, Greulich H, Wong KK. BIBW2992, an irreversible EGFR/HER2 inhibitor highly effective in preclinical lung cancer models. Oncogene 2008; 27(34):4702-4711.

(125) Schutze C, Dorfler A, Eicheler W, Zips D, Hering S, Solca F, Baumann M, Krause M. Combination of EGFR/HER2 tyrosine kinase inhibition by BIBW 2992 and BIBW 2669 with irradiation in FaDu human squamous cell carcinoma. Strahlenther Onkol 2007; 183(5):256-264.

(126) Minkovsky N, Berezov A. BIBW-2992, a dual receptor tyrosine kinase inhibitor for the treatment of solid tumors. Curr Opin Investig Drugs 2008; 9(12):1336-1346.

(127) Rath O, Himmler A, Baum A, Sommergruber W, Beug H, Metz T. c-Myc is required for transformation of FDC-P1 cells by EGFRvIII. FEBS Lett 2007; 581(13):2549-2556.

(128) Pourgholami MH, Morris DL. Inhibitors of vascular endothelial growth factor in cancer. Cardiovasc Hematol Agents Med Chem 2008; 6(4):343-347.

(129) Hilberg F, Roth GJ, Krssak M, Kautschitsch S, Sommergruber W, Tontsch-Grunt U, Garin-Chesa P, Bader G, Zoephel A, Quant J, Heckel A, Rettig WJ. BIBF 1120: triple angiokinase inhibitor with sustained receptor blockade and good antitumor efficacy. Cancer Res 2008; 68(12):4774-4782.

(130) Richardson CJ, Gao Q, Mitsopoulous C, Zvelebil M, Pearl LH, Pearl FM. MoKCa database--mutations of kinases in cancer. Nucleic Acids Res 2009; 37(Database issue):D824-D831.

(131) Shapiro GI. Cyclin-dependent kinase pathways as targets for cancer treatment. J Clin Oncol 2006; 24(11):1770-1783.

(132) Steegmaier M, Hoffmann M, Baum A, Lenart P, Petronczki M, Krssak M, Gurtler U, Garin-Chesa P, Lieb S, Quant J, Grauert M, Adolf GR, Kraut N, Peters JM, Rettig WJ. BI 2536, a potent and selective inhibitor of polo-like kinase 1, inhibits tumor growth in vivo. Curr Biol 2007; 17(4):316-322.

REFERENCES

(133) Bennett MK, Kirk CJ. Development of proteasome inhibitors in oncology and autoimmune diseases. Curr Opin Drug Discov Devel 2008; 11(5):616-625.

(134) Sahin M, Sahin E, Gumuslu S. Cyclooxygenase-2 in Cancer and Angiogenesis. Angiology 2008.

(135) Abbott RG, Forrest S, Pienta KJ. Simulating the hallmarks of cancer. Artif Life 2006; 12(4):617-634.

(136) Garber K. New insights into oncogene addiction found. J Natl Cancer Inst 2007; 99(4):264-5, 269.

(137) Sharma SV, Settleman J. Oncogene addiction: setting the stage for molecularly targeted cancer therapy. Genes Dev 2007; 21(24):3214-3231.

(138) Felsher DW. Oncogene addiction versus oncogene amnesia: perhaps more than just a bad habit? Cancer Res 2008; 68(9):3081-3086.

(139) Weinstein IB, Joe AK. Mechanisms of disease: Oncogene addiction--a rationale for molecular targeting in cancer therapy. Nat Clin Pract Oncol 2006; 3(8):448-457.

(140) Verkman AS, Hara-Chikuma M, Papadopoulos MC. Aquaporins--new players in cancer biology. J Mol Med 2008; 86(5):523-529.

(141) Prevarskaya N, Zhang L, Barritt G. TRP channels in cancer. Biochim Biophys Acta 2007; 1772(8):937-946.

(142) Matthews CP, Colburn NH, Young MR. AP-1 a target for cancer prevention. Curr Cancer Drug Targets 2007; 7(4):317-324.

(143) Myatt SS, Lam EW. The emerging roles of forkhead box (Fox) proteins in cancer. Nat Rev Cancer 2007; 7(11):847-859.

(144) McCarthy PL, Mercer FC, Savicky MW, Carter BA, Paterno GD, Gillespie LL. Changes in subcellular localisation of MI-ER1 alpha, a novel oestrogen receptor-alpha interacting protein, is associated with breast cancer progression. Br J Cancer 2008; 99(4):639-646.

(145) Caldwell GM, Jones CE, Soon Y, Warrack R, Morton DG, Matthews GM. Reorganisation of Wnt-response pathways in colorectal tumorigenesis. Br J Cancer 2008; 98(8):1437-1442.

(146) Arcangeli A, Crociani O, Lastraioli E, Masi A, Pillozzi S, Becchetti A. Targeting ion channels in cancer: a novel frontier in antineoplastic therapy. Curr Med Chem 2009; 16(1):66-93.

(147) Kim BJ, Park EJ, Lee JH, Jeon JH, Kim SJ, So I. Suppression of transient receptor potential melastatin 7 channel induces cell death in gastric cancer. Cancer Sci 2008; 99(12):2502-2509.

REFERENCES

(148) Mao J, Chen L, Xu B, Wang L, Wang W, Li M, Zheng M, Li H, Guo J, Li W, Jacob TJ, Wang L. Volume-activated chloride channels contribute to cell-cycle-dependent regulation of HeLa cell migration. Biochem Pharmacol 2009; 77(2):159-168.

(149) Heo JH, Seo HN, Choe YJ, Kim S, Oh CR, Kim YD, Rhim H, Choo DJ, Kim J, Lee JY. T-type Ca2+ channel blockers suppress the growth of human cancer cells. Bioorg Med Chem Lett 2008; 18(14):3899-3901.

(150) Cianfrocca M, Goldstein LJ. Prognostic and predictive factors in early-stage breast cancer. Oncologist 2004; 9(6):606-616.

(151) Wennmalm K, Calza S, Ploner A, Hall P, Bjohle J, Klaar S, Smeds J, Pawitan Y, Bergh J. Gene expression in 16q is associated with survival and differs between Sorlie breast cancer subtypes. Genes Chromosomes Cancer 2007; 46(1):87-97.

(152) Bergamaschi A, Tagliabue E, Sorlie T, Naume B, Triulzi T, Orlandi R, Russnes HG, Nesland JM, Tammi R, Auvinen P, Kosma VM, Menard S, Borresen-Dale AL. Extracellular matrix signature identifies breast cancer subgroups with different clinical outcome. J Pathol 2008; 214(3):357-367.

(153) Doupnik CA. GPCR-Kir channel signaling complexes: defining rules of engagement. J Recept Signal Transduct Res 2008; 28(1-2):83-91.

(154) Franco R, Ciruela F, Casado V, Cortes A, Canela EI, Mallol J, Agnati LF, Ferre S, Fuxe K, Lluis C. Partners for adenosine A1 receptors. J Mol Neurosci 2005; 26(2-3):221-232.

(155) Sommergruber W, Casari G, Fessl F, Seipelt J, Skern T. The 2A proteinase of human rhinovirus is a zinc containing enzyme. Virology 1994; 204(2):815-818.

(156) Keil B. Specificity of Proteolysis. Springer-Verlag Berlin Heidelberg New York 1992.

(157) Xie J, Shen Z, Li KC, Danthi N. Tumor angiogenic endothelial cell targeting by a novel integrin-targeted nanoparticle. Int J Nanomedicine 2007; 2(3):479-485.

(158) Whittaker CA, Hynes RO. Distribution and evolution of von Willebrand/integrin A domains: widely dispersed domains with roles in cell adhesion and elsewhere. Mol Biol Cell 2002; 13(10):3369-3387.

(159) Dolznig H, Rupp C, Puri C, Sommergruber W, Garin-Chesa P. Developement of a novel *in vitro* 3D-collagen gel co-culture system, which recapitulates tumor-stroma interaction. Jahrestagung der Deutschen Gesellschaft für Pathologie, Berlin 2008; 92.

REFERENCES

(160) Hwang-Verslues WW, Sladek FM. Nuclear receptor hepatocyte nuclear factor 4alpha1 competes with oncoprotein c-Myc for control of the p21/WAF1 promoter. Mol Endocrinol 2008; 22(1):78-90.

(161) Teng SC, Wu KJ, Tseng SF, Wong CW, Kao L. Importin KPNA2, NBS1, DNA repair and tumorigenesis. J Mol Histol 2006; 37(5-7):293-299.

(162) Gluz O, Wild P, Meiler R, Diallo-Danebrock R, Ting E, Mohrmann S, Schuett G, Dahl E, Fuchs T, Herr A, Gaumann A, Frick M, Poremba C, Nitz UA, Hartmann A. Nuclear karyopherin alpha2 expression predicts poor survival in patients with advanced breast cancer irrespective of treatment intensity. Int J Cancer 2008; 123(6):1433-1438.

(163) Sato F, Abraham JM, Yin J, Kan T, Ito T, Mori Y, Hamilton JP, Jin Z, Cheng Y, Paun B, Berki AT, Wang S, Shimada Y, Meltzer SJ. Polo-like kinase and survivin are esophageal tumor-specific promoters. Biochem Biophys Res Commun 2006; 342(2):465-471.

(164) Yu Z, Weinberger PM, Sasaki C, Egleston BL, Speier WF, Haffty B, Kowalski D, Camp R, Rimm D, Vairaktaris E, Burtness B, Psyrri A. Phosphorylation of Akt (Ser473) predicts poor clinical outcome in oropharyngeal squamous cell cancer. Cancer Epidemiol Biomarkers Prev 2007; 16(3):553-558.

(165) Kasid U, Pfeifer A, Brennan T, Beckett M, Weichselbaum RR, Dritschilo A, Mark GE. Effect of antisense c-raf-1 on tumorigenicity and radiation sensitivity of a human squamous carcinoma. Science 1989; 243(4896):1354-1356.

(166) Zhang Q, Thomas SM, Xi S, Smithgall TE, Siegfried JM, Kamens J, Gooding WE, Grandis JR. SRC family kinases mediate epidermal growth factor receptor ligand cleavage, proliferation, and invasion of head and neck cancer cells. Cancer Res 2004; 64(17):6166-6173.

(167) Sheikh Ali MA, Gunduz M, Nagatsuka H, Gunduz E, Cengiz B, Fukushima K, Beder LB, Demircan K, Fujii M, Yamanaka N, Shimizu K, Grenman R, Nagai N. Expression and mutation analysis of epidermal growth factor receptor in head and neck squamous cell carcinoma. Cancer Sci 2008; 99(8):1589-1594.

(168) Dolznig H, Schweifer N, Puri C, Kraut N, Rettig WJ, Kerjaschki D, Garin-Chesa P. Characterization of cancer stroma markers: in silico analysis of an mRNA expression database for fibroblast activation protein and endosialin. Cancer Immun 2005; 5:10.

VDM Verlagsservicegesellschaft mbH

Die VDM Verlagsservicegesellschaft sucht für wissenschaftliche Verlage abgeschlossene und herausragende

Dissertationen, Habilitationen, Diplomarbeiten, Master Theses, Magisterarbeiten usw.

für die kostenlose Publikation als Fachbuch.

Sie verfügen über eine Arbeit, die hohen inhaltlichen und formalen Ansprüchen genügt, und haben Interesse an einer honorarvergüteten Publikation?

Dann senden Sie bitte erste Informationen über sich und Ihre Arbeit per Email an *info@vdm-vsg.de*.

Sie erhalten kurzfristig unser Feedback!

VDM Verlagsservicegesellschaft mbH
Dudweiler Landstr. 99   Telefon +49 681 3720 174
D - 66123 Saarbrücken   Fax   +49 681 3720 1749
www.vdm-vsg.de

Die VDM Verlagsservicegesellschaft mbH vertritt

Printed by Books on Demand GmbH, Norderstedt / Germany